Statistical and Data Handling Skills in Biology

Statistical and Data Handling Skills in Biology

Roland Ennos

PRENTICE HALL

An imprint of PEARSON EDUCATION

Harlow, England · London · New York · Reading, Massachusetts · San Francisco · Toronto · Don Mills, Ontario · Sydney
Tokyo · Singapore · Hong Kong · Seoul · Taipei · Cape Town · Madrid · Mexico City · Amsterdam · Munich · Paris · Milan

Pearson Education Limited
Edinburgh Gate
Harlow
Essex CM20 2JE
England

and Associated Companies around the world

Visit us on the World Wide Web at:
www.pearsoned-ema.com

First published 2000

© Pearson Education Limited 2000

The right of Roland Ennos to be identified as author of
this work has been asserted by him in accordance with
the Copyright, Designs, and Patents Act 1988.

All rights reserved; no part of this publication may be reproduced, stored
in a retrieval system, or transmitted in any form or by any means, electronic,
mechanical, photocopying, recording, or otherwise without either the prior
written permission of the Publishers or a licence permitting restricted copying
in the United Kingdom issued by the Copyright Licensing Agency Ltd,
90 Tottenham Court Road, London W1P OLP.

ISBN 0 582–31278–7

British Library Cataloguing-in-Publication Data
A catalogue record for this book can be obtained from the British Library.

Library of Congress Cataloging-in-Publication Data
A catalog record for this book can be obtained from the Library of Congress.

10 9 8 7 6 5 4 3 2 1
04 03 02 01 00 99

Typeset by 35 in 9½/12pt Concorde BE
Printed and bound in Singapore

ST NO	0582312787
ACC NO	046299
CLASS	570.15195
DATE	21.11.00
[illegible]	AR

To my father

LLYFRGELL COL
BANGOR GW

LLYFRGELL COLEG MENAI LIBRARY
SAFLE FFRIDDOEDD SITE
BANGOR GWYNEDD LL57 2TP

Contents

Preface

Mathematics and statistics are not the most popular subjects of most biology students. This book is designed as a handbook for undergraduates and postgraduates to help them get over this phobia. I hope the information and back-up material provided will give them the confidence to tackle the sort of problems they are likely to encounter in their studies and in future research.

To help put the ideas in context and give students practice, I have included numerous examples and problems. They are mostly concerned with whole-organism biology. This bias reflects my own interests but I hope the simplicity of the examples will make them accessible to a wide range of biologists and help them apply the mathematical ideas to their own particular specialities.

The book is based upon courses given to students at Manchester University School of Biological Sciences. I am heavily indebted to those who took those courses for their feedback. I am also very grateful to my colleague André Gilburn and to Pearson Education's reviewer for pointing out inaccuracies and suggesting improvements to early drafts of the manuscript. I take full responsibility for any errors which remain.

Finally, I would like to thank Yvonne for her support throughout the writing process, even when I was cluttering up the house with manuscripts, computers, books and, most of all, myself.

Roland Ennos

CHAPTER ONE

Introduction

One way of using this book: showing that six minus four equals two

1.1 How to use this book

I believe that biologists only need a small amount of straightforward mathematics to be competent to carry out research. This book aims to help you master these basics with as little pain and distress as possible.

To make the process of learning easier this book takes a strictly biological viewpoint. It tries to explain why certain mathematical concepts can help in biology; it then tries to explain them verbally and using simple graphs, rather than by wholesale use of mathematical equations; and finally, it demonstrates them with the use of numerous worked examples and problems from real biological situations. You will find that the majority of the book is filled up with information on statistical tests. This reflects the great practical importance of statistics in the variable world of biology.

The book has been designed to be used in two ways. You can work through the whole book, either in conjunction with a taught course or on your own. To reinforce the messages and to help you practise your skills, each chapter contains a large number of problems; the solutions are at the end of the book.

Alternatively, you can use the book as a reference manual throughout your academic career. You can recap basic calculations and statistical tests; work out which statistical tests you should use to analyse your experiments; and dip into it to help you design better experiments. To help do this the book contains a range of useful tables and flow charts referring you to the relevant information.

1.2 Finding your way around

1.2.1 Types of quantitative information

There are two very different sorts of quantitative information you can gather when you carry out a survey or perform an experiment: you can take **measurements** from different **samples** or you can count **frequencies** in different **categories**.

1.2.2 Dealing with measurements

The more usual sort of quantitative information is what you gather when you measure things. You might, for instance, be interested in the length, weight, heart rate or age of animals; in the concentration of chemicals in different cells or test tubes; in the pecking order of chickens; or the severity of diseases. Many of these measurements have units and can be combined with other measurements to produce other data. The dimensions of a box, for example, can be combined to work out its volume; or the amount of solute can be combined with the volume of solvent in which it is dissolved to work out its concentration. If you are unsure about the best way to deal with SI units, or how to combine and express measurements, you should read Chapter 2.

Even when you can deal with measurements there is a second problem; all organisms vary, and any data you collect in a biological experiment are bound to be variable. Thus there is no single 'typical value'. Chapter 3 explains why such variability exists, shows how variation can be quantified and expressed, and demonstrates how, by using replication, you can obtain usable information despite the variability.

Of course you will often want to do more than merely take a single set of measurements. You may want to see if the measurements you have taken from a group of organisms you are measuring are different from a particular value and if so by how much. If you are carrying out an experiment, you will probably want to compare measurements from one or more experimentally treated group of organisms or cells with a control group. Similarly, in many surveys you may want to compare measurements from two or more groups of organisms. The aim of such experiments and surveys is to find out whether, and if so by how much, the groups are different. The main problem with doing this is the variability you are bound to have in your measurements. Chapter 4 describes the statistical methods which can help you test whether the **differences** you find are real or whether they just reflect the inevitable variability within the groups.

A further thing you may want to do is to take two or more measurements on a single group of organisms or cells and investigate how the measurements are **associated**. For instance, you may want to investigate how people's heart rates vary with their blood pressure, how weight varies with age, or how the concentrations of different cations in neurones vary with each other. This sort of knowledge can help you work out how organisms work or enable you to predict things about them. Chapter 5 describes some of the ways in which biological measurements might be expected to be related to each other. It goes on to show how these relationships can be represented and identified. Finally, it shows how statistical tests can be used to quantify the relationships between sets of measurements, in spite of their inevitable variability.

1.2.3 Dealing with categorical information

Sometimes you may only be able to collect **categorical** information. This can be gathered when items are divided into a range of different categories. Birds belong to different species and have different colours; bacteria belong to different strains; habitats can have species present or absent; and wind-damaged trees might have been uprooted or had their trunks broken. There is no way you can sensibly measure such arbitrarily defined classes, but what you can do is count the **frequencies** in each category. The ways in which you can deal with these frequencies are described in Chapter 6. This shows how you can determine two things: whether the frequencies of organisms in certain categories are **different** from what you might expect; and whether there are **associations** between categories.

1.2.4 Choosing tests and designing experiments

Finally, the information from all of the early chapters is brought together in Chapter 7. This describes how the skills and background presented in the rest of the book can help you to choose which statistical test to use. It then goes on to describe how you can design and perform better experiments.

LLYFRGELL COLEG MENAI LIBRARY
BANGOR GWYNEDD LL57 2TP

CHAPTER TWO

Dealing with measurements

6.3452 × 10^4, 6.3453 × 10^4..........

2.1 Introduction

It is surprising, considering that most students of biology have studied mathematics for many years, how often they make errors in the ways they deal with and present numerical information. In fact there are many ways of getting things wrong. Primary data can be measured wrongly, or given too high or low a degree of precision. The data can be taken and presented in non-SI units, or mistakes can be made while attempting to convert to SI units. Calculations based on primary data can be carried out incorrectly. Finally, the answers can be given to the wrong degree of precision, in the wrong units, or with no units at all!

Many of the errors are made not only through ignorance but also because of haste, lack of care, or even panic. This chapter shows how you can avoid such mistakes by carrying out the following logical sequence of steps carefully and in the right order: measuring, converting data into SI units, combining data together, and expressing the answer in SI units to the correct degree of precision.

2.2 Measuring

Measurements should always be taken to the highest possible degree of precision. This is straightforward with modern digital devices, but it is more difficult in the more old-fashioned devices, which have a graduated analogue scale. The highest degree of precision of analogue instruments is usually to the smallest graduation of the scale. Using a 30 cm ruler, lengths can only be measured to the nearest millimetre. However, if the graduations are far enough apart, as they are on some thermometers, it is usually possible to judge the measurements to the next decimal place. This is made even easier by devices, like calipers or microscope stages, which have a vernier scale.

2.3 Converting to SI units

2.3.1 SI units

Before carrying out any further manipulation of data or expressing it, it should be converted into the correct SI units. The *Système International d'Unités* (SI) is the accepted scientific convention for measuring physical quantities, under which the most basic units of length, mass and time are kilograms, metres and seconds respectively. The complete list of the basic SI units is given in Table 2.1.

All other units are derived from these basic units. For instance, volume should be expressed in cubic metres or m^3. Similarly density, mass per unit volume, should be expressed in kilograms per cubic metre or $kg\ m^{-3}$. Some important derived units have their own names; the unit of force ($kg\ m\ s^{-2}$) is called a newton (N), and the unit of pressure ($N\ m^{-2}$) is called a pascal (Pa). A list of important derived SI units is given in Table 2.2.

Table 2.1 The base and supplementary SI units.

Measured quantity	*SI unit*	*Symbol*
Base		
Length	metre	m
Mass	kilogram	kg
Time	second	s
Amount of substance	mole	mol
Temperature	kelvin	K
Electric current	ampere	A
Luminous intensity	candela	cd
Supplementary		
Plane angle	radian	rad
Solid angle	steradian	sr

Table 2.2 Important derived SI units.

Measured quantity	*Name of unit*	*Symbol*	*Definitions*
Mechanics			
Force	newton	N	kg m s^{-2}
Energy	joule	J	N m
Power	watt	W	J s^{-1}
Pressure	pascal	Pa	N m^{-2}
Electricity			
Charge	coulomb	C	A s
Potential difference	volt	V	J C^{-1}
Resistance	ohm	Ω	V A^{-1}
Conductance	siemens	S	Ω^{-1}
Capacitance	farad	F	C V^{-1}
Light			
Luminous flux	lumen	lm	cd sr^{-1}
Illumination	lux	lx	lm m^{-2}
Others			
Frequency	hertz	Hz	s^{-1}
Radioactivity	becquerel	Bq	s^{-1}
Enzyme activity	katal	kat	mol substrate s^{-1}

2.3.2 Dealing with large and small numbers

The problem with using a standard system, like the SI system, is that the units may not always be convenient. The mass of organisms ranges from 0.000 000 000 1 kg for algae to 100 000 kg for whales. For convenience, therefore, two different systems can be used to present large and small measurements. Both these systems also have the added advantage that large numbers can be written without using a large number of zeros, which would imply an unrealistic degree of precision. It would be difficult to weigh a whale to the nearest kilogram (and pointless, since the weight will fluctuate wildly at this degree of precision), which is what the weight of 100 000 kg implies.

Use of prefixes

Using prefixes, each of which stands for a multiplication factor of 1000 (Table 2.3), is the simplest way to present large or small measurements. Any quantity can be simply presented as a number between 0.1 and 1000 multiplied by a suitable prefix. For instance, 123 000 J is better presented as 123 kJ or 0.123 MJ. Similarly, 0.000 012 m is better presented as 12 μm (not 0.012 mm).

Use of scientific notation

The problem with using prefixes is that they are rather tricky to combine mathematically when carrying out calculations. For this reason, when

Table 2.3 Prefixes used in SI.

Small numbers						
Multiple	10^{-3}	10^{-6}	10^{-9}	10^{-12}	10^{-15}	10^{-18}
Prefix	milli	micro	nano	pico	femto	atto
Symbol	m	μ	n	p	f	a
Large numbers						
Multiple	10^{3}	10^{6}	10^{9}	10^{12}	10^{15}	10^{18}
Prefix	kilo	mega	giga	tera	peta	exa
Symbol	k	M	G	T	P	E

performing a calculation it is usually better to express your data using **scientific notation**. As we shall see, this makes calculations much easier.

Any quantity can be expressed as a number between 1 and 10 multiplied by a power of 10 (also called an exponent). For instance, 123 is equal to 1.23 multiplied by 10 squared or 10^2. Here the exponent is 2, so it can be written as 1.23×10^2. Similarly, 0.001 23 is equal to 1.23 multiplied by the inverse of 10 cubed, or 10^{-3}. Therefore it is best written as 1.23×10^{-3}. And 1.23 itself is equal to 1.23 multiplied by 10 to the power 0, so it does not need an exponent.

A simple way of determining the value of the exponent is to count the number of digits from the decimal point to the right of the first significant figure. For instance, in 18 000 there are four figures to the right of the 1, which is the first significant figure, so $18\,000 = 1.8 \times 10^4$. Similarly, in 0.000 000 18 there are seven figures between the point and the right of the 1, so $0.000\,000\,18 = 1.8 \times 10^{-7}$.

Prefixes can readily be converted to exponents, since each prefix differs by a factor of 1000 or 10^3 (Table 2.3). The pressure 4.6 MPa equals 4.6×10^6 Pa, and 46 MPa equals $4.6 \times 10^1 \times 10^6 = 4.63 \times 10^7$ Pa.

2.3.3 Converting from non-SI units

Very often textbooks and papers, especially old ones, present quantities in non-SI units, and old apparatus may also be calibrated in non-SI units. Before carrying out calculations, you will need to convert them to SI units. Fortunately, this is very straightforward.

Non-SI metric units

The most common non-SI units are those which are metric but based on obsolete systems. The most useful biological examples are given in Table 2.4, along with a conversion factor. These units are very easy to convert into the SI system. Simply multiply your measurement by the conversion factor.

Table 2.4 Conversion factors from obsolete units to SI.[a]

Quantity	*Old unit/Symbol*	*SI unit/Symbol*	*Conversion factor*
Length	*angstrom/Å*	metre/m	1×10^{-10}
	yard	metre/m	0.9144
	foot	metre/m	0.3048
	inch	metre/m	2.54×10^{-2}
Area	*hectare/ha*	square metre/m^2	1×10^{4}
	acre	square metre/m^2	4.047×10^{3}
	square foot/ft^2	square metre/m^2	9.290×10^{-2}
	square inch/in^2	square metre/m^2	6.452×10^{-4}
Volume	*litre/l*	cubic metre/m^3	1×10^{-3}
	cubic foot/ft^3	cubic metre/m^3	2.832×10^{-2}
	cubic inch/in^3	cubic metre/m^3	1.639×10^{-5}
	UK pint/pt	cubic metre/m^3	5.683×10^{-4}
	US pint/liq pt	cubic metre/m^3	4.732×10^{-4}
	UK gallon/gal	cubic metre/m^3	4.546×10^{-3}
	US gallon/gal	cubic metre/m^3	3.785×10^{-3}
Angle	degree/°	radian/rad	1.745×10^{-2}
Mass	*tonne*	kilogram/kg	1×10^{3}
	ton (UK)	kilogram/kg	1.016×10^{3}
	hundredweight/cwt	kilogram/kg	5.080×10^{1}
	stone	kilogram/kg	6.350
	pound/lb	kilogram/kg	0.454
	ounce/oz	kilogram/kg	2.835×10^{-2}
Energy	*erg*	joule/J	1×10^{-7}
	kilowatt hour/kWh	joule/J	3.6×10^{6}
Pressure	*bar/b*	pascal/Pa	1×10^{5}
	mm Hg	pascal/Pa	1.332×10^{2}
Radioactivity	curie/Ci	becquerel/Bq	3.7×10^{10}
Temperature	*centigrade/°C*	kelvin/K	$C + 273.15$
	Fahrenheit/°F	kelvin/K	$\frac{5}{9}(F + 459.7)$

[a] Metric units are given in italics. To get from a measurement in the old unit to a measurement in the SI unit, multiply by the conversion factor.

Example 2.1

Give the following in SI units:

(a) 24 ha

(b) 25 cm

Solution

(a) 24 ha equals 24×10^4 m^2 = 2.4×10^5 m^2

(b) 25 cm equals 25×10^{-2} m = 2.5×10^{-1} m

Litres and concentrations

The most important example of a unit which is still widely used even though it does not fit into the SI system is the litre (1 dm^3 or 10^{-3} m^3), which is used in the derivation of the concentration of solutions. For instance, if 1 litre contains 2 moles of a substance then its concentration is given as 2 *M* or molar.

The mole is now a bona fide SI unit, but it too was derived before the SI system was developed, since it was originally the amount of a substance which contains the same number of particles as 1 g (rather than the SI kilogram) of hydrogen atoms. In other words, the mass of 1 mole of a substance is its molecular mass in grams.

When working out concentrations of solutions it is probably best to stick to these units, since most glassware is still calibrated in litres and small balances in grams.

The molarity *M* of a solution is obtained as follows:

$$M = \frac{\text{Number of moles}}{\text{Solution volume (l)}}$$

$$= \frac{\text{Mass (g)}}{\text{Molecular mass} \times \text{Solution volume (l)}}$$

Example 2.2

A solution contains 23 g of copper sulphate ($CuSO_4$) in 2.5 litres of water. What is its concentration?

Solution

$$\text{Concentration} = 23/((63.5 + 32 + 64) \times 2.5)$$

$$= 5.768 \times 10^{-2}\ M$$

$$= 5.8 \times 10^{-2}\ M \quad \text{(2 significant figures)}$$

Non-metric units

Non-metric units, such as those based on the old Imperial scale, are also given in Table 2.4. Again you must simply multiply your measurement by the conversion factor. However, they are more difficult to convert to SI, since they must be multiplied by factors which are not just powers of 10. For instance,

$$6 \text{ ft} = 6 \times 3.048 \times 10^{-1} \text{ m} = 1.83 \text{ m}$$

Note that the answer was given as 1.83 m, not the calculated figure of 1.8288 m. This is because a measure of 6 ft implies that the length was measured to the nearest inch. The answer we produced is accurate to the nearest centimetre, which is the closest SI unit.

If you have to convert square or cubic measures into metric, simply multiply by the conversion factor to the power of 2 or 3. So 12 cubic feet $= 12 \times (3.038 \times 10^{-1})^3 \text{ m}^3 = 3.4 \times 10^{-1} \text{ m}^3$ to 2 significant figures.

2.4 Combining values

Once measurements have been converted into SI units with exponents, they are extremely straightforward to combine using either pencil and paper or calculator (most calculators use exponents nowadays). When multiplying two measurements, for instance, you simply multiply the initial numbers, add the exponents and multiply the units together. If the multiple of the two initial numbers is greater than 10 or less than 1, you simply add or subtract 1 from the exponent. For instance,

$$\begin{aligned} 2.3 \times 10^2 \text{ m} \times 1.6 \times 10^3 \text{ m} &= (2.3 \times 1.6) \times 10^{(2+3)} \text{ m}^2 \\ &= 3.7 \times 10^5 \text{ m}^2 \end{aligned}$$

Notice that the area is given to 2 significant figures because that was the degree of precision with which the lengths were measured. Similarly,

$$\begin{aligned} 2.3 \times 10^2 \text{ m} \times 6.3 \times 10^{-4} \text{ m} &= (2.3 \times 6.3) \times 10^{(2-4)} \text{ m}^2 \\ &= 1.4 \times 10^1 \times 10^{-2} \text{ m}^2 \\ &= 1.4 \times 10^{-1} \text{ m}^2 \end{aligned}$$

In the same way, when dividing one measurement by another you divide the first initial number by the second, subtract the second exponent from the first, and divide the first unit by the second.

$$\begin{aligned} \text{Therefore} \quad (4.8 \times 10^3 \text{ m}) \ / \ (1.5 \times 10^2 \text{ s}) &= (4.8/1.5) \times 10^{(3-2)} \text{ m s}^{-1} \\ &= 3.2 \times 10^1 \text{ m s}^{-1} \end{aligned}$$

2.5 Expressing the answer

When you have completed all calculations, you must be careful how you express your answer. First, it should be given to the same level of precision as the *least* accurate of the measurements from which it was calculated. This book and many statistical packages use the following convention: the digits 1 to 4 go down, 6 to 9 go up and 5 goes to the nearest even digit. Here are some examples:

0.343	becomes	0.34	to	2	significant figures
0.2251	becomes	0.22	to	2	significant figures
0.6354	becomes	0.64	to	2	significant figures

Second, it is sometimes a good idea to express it using a prefix. So if we work out from figures given to two significant figures that a pressure is 2.678×10^6 Pa, it should be expressed as 2.7 MPa. Always adjust the degree of precision *at the end* of the calculation.

2.6 Doing all three steps

The various steps can now be carried out to manipulate data to reliably derive further information. It is important to carry out each step in its turn, producing an answer before going on to the next step in the calculation. Doing all the calculations at once can cause confusion and lead to silly mistakes.

Example 2.3

A sample of heartwood taken from an oak tree was 12.1 mm long by 8.2 mm wide by 9.5 mm deep and had a wet mass of 0.653 g. What was its density?

Solution

Density has units of mass (in kg) per unit volume (in m^3). Therefore the first thing to do is to convert the units into kg and m. The next thing to do is to calculate the volume in m^3. Only then can the final calculation be performed. This slow building up of the calculation is ponderous but is the best way to avoid making mistakes.

$$\text{Mass} = 6.53 \times 10^{-4} \text{ kg}$$

$$\text{Volume} = 1.21 \times 10^{-2} \times 8.2 \times 10^{-3} \times 9.5 \times 10^{-3}$$

$$= 9.4259 \times 10^{-7} \text{ m}^3$$

$$\text{Density} = \frac{\text{mass}}{\text{volume}} = \frac{6.53 \times 10^{-4}}{9.4259 \times 10^{-7}}$$

$$= 0.6928 \times 10^3 \text{ kg m}^{-3}$$

$$= 6.9 \times 10^2 \text{ kg m}^{-3}$$

Notice that the answer is given to two significant figures, like the dimensions of the sample.

LLYFRGELL COLEG MEN[illegible]
BANGOR GWYNEDD [illegible]

Table 2.5 Some useful constants and formulae.

Physical constants	
Density of water	$= 1000\ \text{kg m}^{-3}$
Density of air	$= 1.2\ \text{kg m}^{-3}$
Specific heat of water	$= 4.2 \times 10^3\ \text{J K}^{-1}\ \text{kg}^{-1}$
Chemical constants	
1 mol	$= 6 \times 10^{23}$ particles
Mass of 1 mol	= molecular mass (g) $= 10^{-3} \times$ molecular mass (kg)
Volume of 1 mol of Gas	$= 24\ \text{l} = 2.4 \times 10^{-2}\ \text{m}^3$ (at room temperature and pressure)
1 molar solution (1 *M*)	$= 1\ \text{mol l}^{-1} = 1000\ \text{mol m}^{-3}$
1 normal Solution (1 *N*)	$= 1\ \text{mol l}^{-1} = 1000\ \text{mol m}^{-3}$ of ions
$\text{pH} = -\log_{10}[\text{H}^+]$	
Composition of air	= 78.1% nitrogen, 20.9% oxygen, 0.93% argon and 0.03% carbon dioxide, plus traces of others, by volume
Mathematical formulae	
Area of a circle of radius *R*	$= \pi R^2$
Volume of a sphere of radius *R*	$= \frac{4}{3}\pi R^3$
Area of a sphere of radius *R*	$= 4\pi R^2$
Volume of a cylinder of radius *R* and height *H*	$= \pi R^2 H$
Volume of a cone of radius *R* and height *H*	$= \frac{1}{3}\pi R^2 H$
Mathematical constants	
$\pi = 3.1416$	
$\log_e x = 2.30 \log_{10} x$	

2.7 Constants and formulae

Frequently, raw data on their own are not enough to work out other important quantities. You may need to include physical or chemical constants in your calculations, or insert your data into basic mathematical formulae. Table 2.5 is a list of some useful constants and formulae. Many of them are worth memorising.

Example 2.4

A total of 25 micropropagated plants were grown in a 10 cm diameter Petri dish. At what density were they growing?

Solution

The first thing to calculate is the area of the Petri dish. Since its diameter is 10 cm, its radius R will be 5 cm (or 5×10^{-2} m). A circle's area A is given by the formula $A = \pi R^2$. Therefore

$$\text{Area} = 3.1416 \times (5 \times 10^{-2})^2$$
$$= 7.854 \times 10^{-3}\ \text{m}^2$$

The density is the number per unit area, so

$$\text{Density} = 25/(7.854 \times 10^{-3})$$
$$= 3.183 \times 10^{3}\ \text{m}^{-2}$$
$$= 3.2 \times 10^{3}\ \text{m}^{-2} \quad \text{(2 significant figures)}$$

2.8 Using calculations

Once you can reliably perform calculations, you can use them for far more than just working out the results of your experiments from your raw data. You can use them to put your results into perspective or extrapolate from your results into a wider context. You can also use calculations to help design your experiments: to work out how much of each ingredient you need, or how much the experiment will cost. But even more usefully, they can help you to work out whether a particular experiment is worth attempting in the first place. Calculations are thus an invaluable tool for the research biologist to help save time, money and effort. They don't even have to be very exact calculations. Often, all that is required is to work out a rough or ball-park figure.

Example 2.5

Elephants are the most practical form of transport through the Indian rainforest because of the rough terrain; the only disadvantage is their great weight. A scientific expedition needs to cross a bridge with a weight limit of 10 tonnes, in order to enter a nature reserve. Will their elephants be able to cross this bridge safely?

Solution

You are unlikely, in the rainforest, to be able to look up or measure the weight of an elephant, but most people have some idea of just how big they are. Since the mass of an object is equal to volume × density, the first thing to calculate is the volume. What is the volume of an elephant? Well, elephants are around 2–3 m long and have a (very roughly) cylindrical body of diameter, say, 1.5 m (so the radius = 0.75 m). The volume of a cylinder is given by $V = \pi R^2 L$, so with these figures the volume of the elephant is approximately

$$V = \pi \times 0.75^2 \times 2 \quad \text{up to} \quad \pi \times 0.75^2 \times 3$$

$$V = 3.53\text{–}5.30 \text{ m}^3$$

The volume of the legs, trunk, etc., is very much less and can be ignored in this rough calculation. So what is the density of an elephant? Well, elephants (like us) can just about float in water and certainly swim, so they must have about the same density as water, 1000 kg m^{-3}. The approximate mass of the elephant is therefore

$$\text{Mass} = 1000 \times (3.5 \text{ to } 5.3)$$

$$= 3530\text{–}5300 \text{ kg}$$

Note, however, that the length of the beast was estimated to only one significant figure, so the weight should also be estimated to this low degree of accuracy. The weight of the elephant will be (4–5) $\times 10^3$ kg or 4–5 tonnes. (Textbook figures for weights of elephants range from 3 to 7 tonnes.) The bridge should easily be able to withstand the weight of an elephant.

This calculation would not have been accurate enough to determine whether our elephant could cross a bridge with weight limit 4.5 tonnes. It would have been necessary to devise a method of weighing it.

2.9 Logarithms, graphs and pH

2.9.1 Logarithms to base 10

Though scientific notation (such as 2.3×10^4) is a good way of expressing large and small numbers (such as 23 000), it is a bit clumsy since the numbers consist of two parts, the initial number and the exponent. Nor does it help very large and very small numbers to be conveniently represented on the same graph. For instance, if you plot the relationship between the numbers of bird species in woods of areas 100, 1000, 10 000, 100 000 and 1 000 000 m^2 (Figure 2.1a), most of the points will be congested at the left.

These problems can be overcome by the use of **logarithms**. Any number can be expressed as a single **exponent**, as 10 to the power of a second number, e.g. $23\,000 = 10^{4.362}$ The 'second number' (here 4.362) is called the logarithm to base 10 ($\log_{10}$) of the first, so that

$$4.362 = \log_{10} 23\,000$$

Numbers above 1 have a positive logarithm, whereas numbers below 1 have a negative logarithm, e.g.

$$0.0045 = 10^{-2.347} \quad \text{so} \quad -2.347 = \log_{10} 0.0045$$

Logarithms to the base 10 of any number can be found simply by pressing the log button on your calculator, and can be converted back to real numbers by pressing the 10^x button on your calculator.

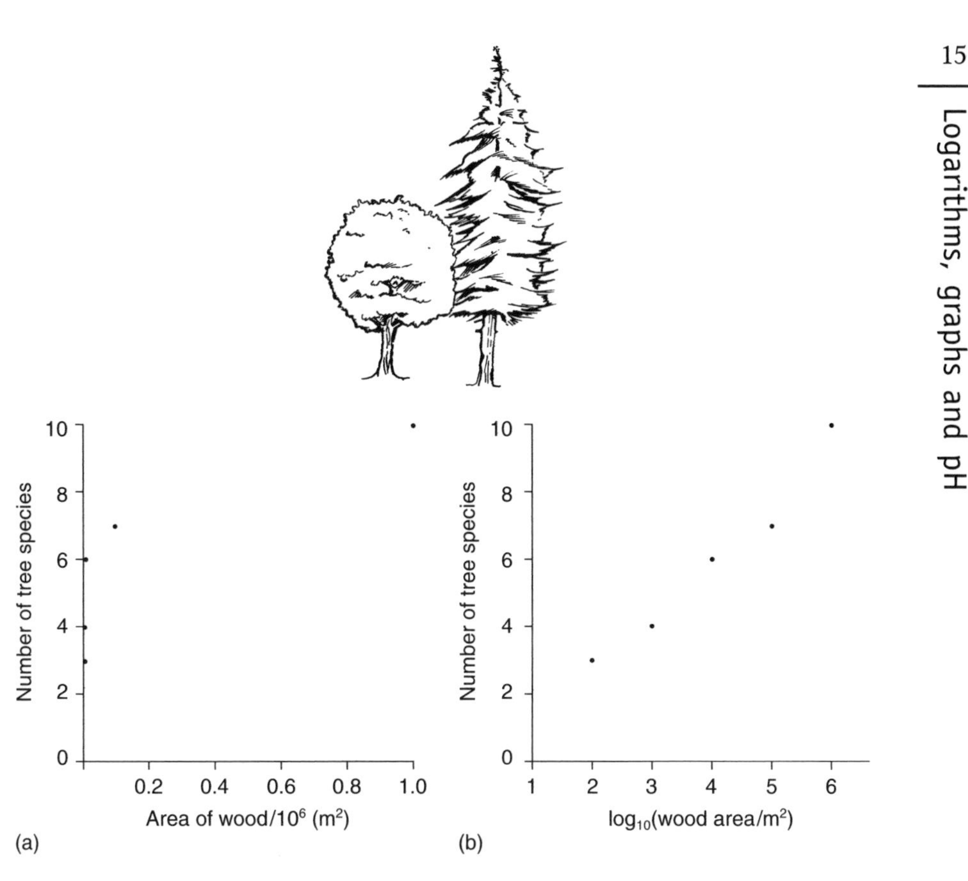

Figure 2.1 Using logarithms. (a) A simple graph showing the relationship between the size of woods and the number of tree species they contain; the points are hopelessly congested at the left of the plot. (b) Plotting number of species against $\log_{10}$ (area) spreads the data out more evenly.

Properties and uses of logarithms

The most important property of logarithms is that if numbers have a constant ratio between them, their logarithms will differ by a constant amount. Hence the numbers 1, 10 and 100, which differ by ratios of 10, have logarithms of 0, 1 and 2, which differ by 1 each time. This gives them some useful mathematical properties, which can help us work out relationships between variables, as we shall see in Chapter 5. However, it also gives them two immediate uses.

Use of logarithms for graphs

Logarithms allow very different quantities to be compared and plotted on the same graph. For instance, you can show the relationship between wood area and number of tree species (Figure 2.1a) more clearly by plotting species number against $\log_{10}$ (area) (Figure 2.1b).

pH

The single most important use of logarithms in biology is in the units for acidity. The unit pH is given by the formula

$$\mathrm{pH} = -\log_{10}[\mathrm{H}^+] \qquad (2.1)$$

where $[\mathrm{H}^+]$ is the hydrogen ion concentration in moles per litre (M). Therefore, a solution containing 2×10^{-5} mole (mol) of hydrogen ions per litre will have a pH of $-\log_{10}(2 \times 10^{-5}) = 4.7$.

Example 2.6

A solution has a pH of 3.2. What is the hydrogen ion concentration?

Solution

The hydrogen ion concentration is $10^{-3.2} = 6.3 \times 10^{-4}$ M.

2.9.2 Natural logarithms

Logarithms can be calculated for other bases as well as 10. Other important types of logarithms are **natural logarithms** ($\log_e$ or ln) in which numbers that differ by the ratio 2.718 (which is given the letter e) have logs that differ by 1. Thus ln 2.718 = 1. As we shall see in Chapter 5, natural logarithms are particularly useful when describing and investigating exponential increases in populations or exponential decay in radioactivity.

To convert from a number to its natural logarithm, you should press the ln button on your calculator. To convert back, you should press the e^x button.

2.10 Self-assessment problems

Problem 2.1

What are the SI units for the following measurements?

(a) Area
(b) The rate of height growth for a plant
(c) The concentration of red cells in blood
(d) The ratio of the concentrations of white and red cells in blood

Problem 2.2

How would you express the following quantities using appropriate prefixes?

(a) 192 000 000 N
(b) 0.000 000 102 kg

(c) 0.000 12 s
(d) 21.3 cm

Problem 2.3

How would you express the following quantities in scientific notation using appropriate exponents?

(a) 0.000 046 1 J
(b) 461 000 000 s

Problem 2.4

How would you express the following quantities in scientific notation using the appropriate exponents?

(a) 3.81 GPa
(b) 4.53 mW
(c) 364 mJ
(d) 4.8 mg
(e) 0.21 pg

Problem 2.5

Convert the following to SI units expressed in scientific notation.

(a) 250 tonnes
(b) 0.3 bar
(c) 24 angstroms

Problem 2.6

Convert the following into SI units.

(a) 35 yards
(b) 3 feet 3 inches
(c) 9.5 square yards

Problem 2.7

Perform the following calculations.

(a) 1.23×10^3 m $\times 2.456 \times 10^5$ m
(b) $(2.1 \times 10^{-2}$ J) / $(4.5 \times 10^{-4}$ kg)

Problem 2.8

Give the following expressions in prefix form and to the correct degree of precision.

(a) 1.28×10^{-3} mol to 2 significant figures
(b) 3.649×10^{8} J to 3 significant figures
(c) 2.423×10^{-7} m to 2 significant figures

Problem 2.9

A blood cell count was performed. Within the box on the slide, which had sides of length 1 mm and depth of 100 μm, there were 652 red blood cells. What was the concentration of cells (in m^{-3}) in the blood?

Problem 2.10

An old-fashioned rain gauge showed that 0.6 inches of rain had fallen on an experimental plot of area of 2.6 ha. What volume of water had fallen on the area?

Problem 2.11

What is the concentration of a solution of 25 g of glucose (formula $C_6H_{12}O_6$) in a volume of 2000 ml of water?

Problem 2.12

An experiment to investigate the basal metabolic rate of human beings showed that in 5 minutes the subject breathed out 45 litres of air into a Douglas bag. The oxygen concentration in this air had fallen from 19.6% by volume to 16.0%, so it contained 3.6% CO_2 by volume. What was the mass of this CO_2 and at what rate had it been produced?

Problem 2.13

A chemical reaction heated 0.53 litre of water by 2.4 K. How much energy had it produced?

Problem 2.14

An experiment which must be repeated around 8 times requires 80 ml of a 3×10^{-3} *M* solution of the substance X. Given that X has a molecular mass of 258 and costs £56 per gram, and given that your budget for the year is £2000, do you think you will be able to afford to do the experiment?

Problem 2.15

It has been postulated that raised bogs may be major producers of methane and, because methane is a greenhouse gas, therefore an important cause of the greenhouse effect. A small microcosm experiment was carried out to investigate the rate at which methane is produced by a raised bog in North Wales. This showed that the rate of production was 21 ml m^{-2} per day. Given that (1) world production of CO_2 by burning fossil fuels is 25 Gt per year, (2) weight for weight methane is said to be 3 times more efficient a greenhouse gas than CO_2 and (3) there is 3.4×10^6 km^2 of blanket bog in the world, what do you think of this idea?

Problem 2.16

Calculate $\log_{10}$ of

(a) 45
(b) 450
(c) 0.000 45
(d) 1 000 000
(e) 1

Problem 2.17

Reconvert the following logarithms to base 10 back to numbers.

(a) 1.4
(b) 2.4
(c) −3.4
(d) 4
(e) 0

Problem 2.18

Calculate the pH of the following solutions.

(a) 3×10^{-4} *M* HCl
(b) 4×10^{-6} *M* H_2SO_4

Problem 2.19

Calculate the mass of sulphuric acid (H_2SO_4) in 160 ml of a solution which has a pH of 2.1.

Problem 2.20

Calculate the natural logarithm of

(a) 30
(b) 0.024
(c) 1

Problem 2.21

Convert the following natural logarithms back to numbers.

(a) 3
(b) –3
(c) 0

CHAPTER THREE

Dealing with variation

So men aren't all the same!

3.1 Introduction

We saw in the last chapter that you have to be extremely careful in the way you take measurements, manipulate them and express them to produce reliable results. However, on its own that is not enough. One of the main problems biologists encounter when they carry out research is that because all organisms are unique no useful information can be gleaned from just a single measurement. For instance, if you have reliable scales and want to measure exactly how much bull elephants weigh, it is no use just weighing one animal. Because of the great deal of variation seen in nature, we would not know how characteristic it was of elephants as a whole.

It is because of the problem of variation that we need to do so much work in biology carrying out **replicated** surveys and experiments. In turn, we need to analyse the results using complex statistical tests to tease out the variation.

This chapter outlines how and why biological measurements vary, describes how variation is quantified, and finally shows how, by combining results from several measurements, you can obtain useful quantitative information despite the variation.

3.2 Variability: causes and effects

There are three main reasons why the measurements we take of biological phenomena vary. The first is that organisms differ because their genetic make-up varies. Most of the continuous characters, like height, weight, metabolic rate or blood [Na^+], are influenced by a large number of genes, each of which has a small effect; they act to either increase or decrease the value of the character by a small amount. Second, organisms also vary because they are influenced by a large number of environmental factors, each of which has similarly small effects. Third, we may make a number of small errors in our actual measurements.

So how will these factors influence the **distribution** of the measurements we take? Let's look first at the simplest possible system; imagine a population of rats whose length is influenced by a single factor found in two forms. Half the time it is found in the form which increases length by 20% and half the time in the form which decreases it by 20%. The distribution of heights will be that shown in Figure 3.1a. Half the rats will be 80% of the average length and half 120% of the average length.

What about the slightly more complex case in which length is influenced by two factors, each of which is found half the time in a form which increases length by 10% and half the time in a form which decreases it by 10%. Of the four possible combinations of factors, there is one in which both factors increase length (and hence length will be 120% of average), and one in which they both reduce length (making length 80% of average). The chances of being either long or short are $\frac{1}{2} \times \frac{1}{2} = \frac{1}{4}$. However, there are two possible cases in which overall length is average: if the first factor increases length and the second decreases it; and if the first factor decreases length and the second increases it. Therefore one-half of the rats will have average length (Figure 3.1b).

Figure 3.1c gives the results for the even more complex case when length is influenced by four factors, each of which is found half the time in the form which increases length by 5% and half the time in the form which decreases it by 5%. In this case, of 16 possible combinations of factors, there is only one combination in which all four factors are in the long form and one combination in which all are in the short form. The chances of each are therefore $\frac{1}{2} \times \frac{1}{2} \times \frac{1}{2} \times \frac{1}{2} = \frac{1}{16}$. The rats are much more likely to be intermediate in size, because there are four possible combinations in which three long and one short factor (or three short and one long) can be arranged, and six possible combinations in which two long and two short factors can be arranged. It can be seen that the central peak is higher than those further out. The process is even more apparent, and the shape of the distribution becomes more obviously humped if there are eight factors, each of which increases or decreases length by 2.5% (Figure 3.1d). The resulting distributions are known as **binomial distributions**.

If length were affected by more and more factors, this process would continue; the curve would become smoother and smoother until, if length were affected by an infinite number of factors, we would get a bowler-hat-

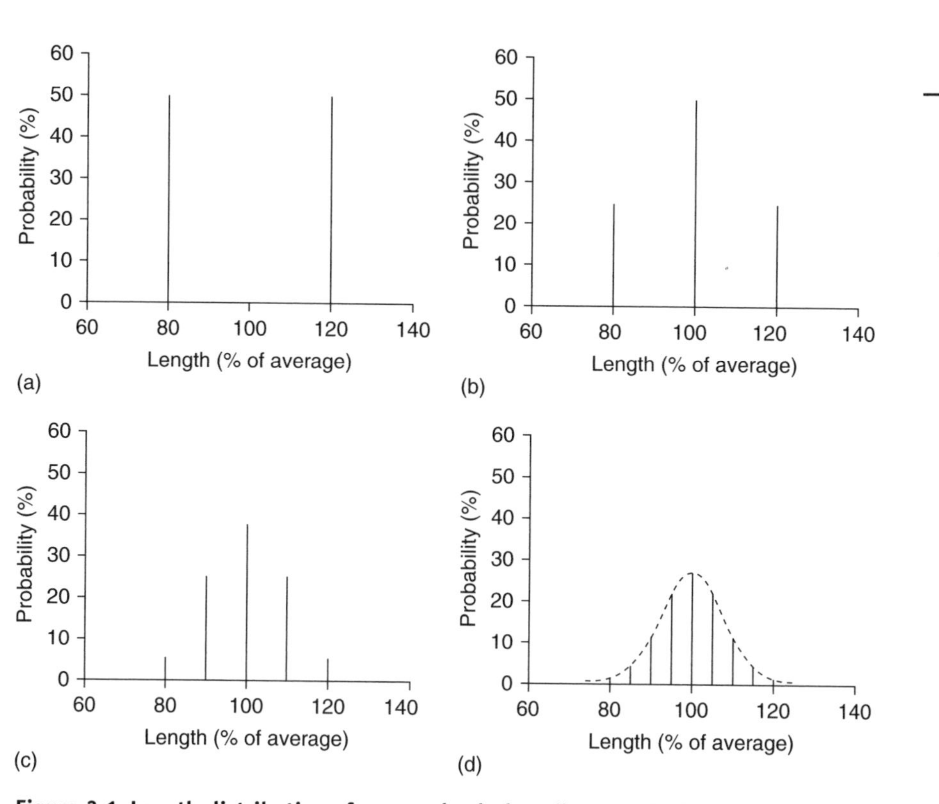

Figure 3.1 Length distributions for a randomly breeding population of rats. Length is controlled by a number of factors, each of which is found 50% of the time in the form which reduces length and 50% in the form which increases length. The graphs show length control by (a) 1 factor, (b) 2 factors, (c) 4 factors and (d) 8 factors. The greater the number of influencing factors, the greater the number of peaks and the more nearly they approximate a smooth curve (dashed outline).

shaped distribution curve (Figure 3.2). This is the so-called **normal distribution** (also known as the Z distribution). If we measured an infinite number of rats, most would have length somewhere near the average, and the numbers would tail off on each side.

3.3 Describing the normal distribution

Once we know (or assume) that a measurement follows the normal distribution, we can describe the distribution of the measurements using just two numbers. The position of the centre of the distribution is described by the **population mean** μ, which on the graph is located at the central peak of the distribution. The mean is the average value of the measurement and is found mathematically by dividing the sum of the lengths of all the rats by the number of rats:

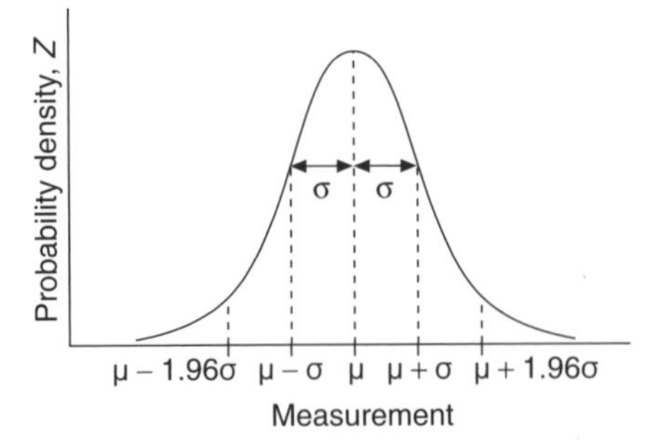

Figure 3.2 A normal distribution. Here 68% of measurements are found within one standard deviation σ from the mean μ; 95% are found within 1.96 times the standard deviation from the mean.

$$\mu = \frac{\Sigma x_i}{N} \tag{3.1}$$

where x_i is the values of length and N is the number of rats.

The width of the distribution is described by the **population standard deviation** σ, which is the distance from the central peak to the point of inflexion of the curve (where it changes from being convex to concave). This standard deviation is a measure of about how much, on average, points differ from the mean. It is actually calculated by a two-stage process. The first stage is to calculate the **population variance** V, which is the average amount of the square of the distance of each point from the mean. The variance is therefore equal to the 'sum of squares' divided by the number of points:

$$V = \frac{\Sigma(x_i - \mu)^2}{N} \tag{3.2}$$

To calculate the **standard deviation** it is necessary to take the square root of this value, which gets us back to the same units as the mean. Mathematically, standard deviation σ is given by

$$\sigma = \sqrt{\frac{\Sigma(x_i - \mu)^2}{N}} \tag{3.3}$$

It turns out that 68.2% of measurements will be within one standard deviation of the mean, 95% of all measurements will be within 1.96 times the standard deviation from the mean, 99% within 2.58 times the standard deviation from the mean and 99.9% within 3.29 times the standard deviation from the mean.

Example 3.1

Suppose adult cats have a mean mass of 3.52 kg with a standard deviation of 0.65 kg. What are the upper and lower limits of mass between which 95% of the cats are found?

Solution

We know that 95% of cats will be within $(1.96 \times 0.65) = 1.27$ kg of 3.52 kg. Therefore 95% will have mass between 2.25 kg and 4.79 kg.

3.4 Estimating the mean and standard deviation

It is all very well being able to say things about populations whose means and standard deviations we know with certainty. However, in real life it is virtually impossible to find the exact mean and standard deviation of any population. To calculate them we would have to take an infinite number of measurements!

The only practical thing to do is to take a **sample** of a manageable size and use the results from the measurements we have taken to **estimate** the population mean and standard deviation. It is very easy to calculate an **estimate of the population mean**. It is simply the average of the sample, or the sample mean $\bar{x}$. This is calculated just like the population mean; it is simply the sum of all the lengths divided by the number of rats measured. In mathematical terms this is given by the expression

$$\bar{x} = \frac{\Sigma x_i}{N} \tag{3.4}$$

where x_i is the values of length and N is the number of rats.

The **estimate of the population standard deviation**, written s or σ_{n-1}, is given by a different expression from equation (3.3). Rather than dividing the sum of squares by N, we divide by $(N-1)$ to give the formula

$$s = \sigma_{n-1} = \sqrt{\frac{\Sigma(x_i - \bar{x})^2}{N-1}} \tag{3.5}$$

We use $(N-1)$ because this expression will give an unbiased estimate of the population standard deviation, whereas using N would tend to underestimate it. To see why this is so, it is perhaps best to consider the case when we have only taken one measurement. Since the estimated mean $\bar{x}$ necessarily equals the single measurement, the standard deviation we calculate when we use N will be zero. Similarly, if there are two points, the estimated mean will be constrained to be exactly halfway between them, whereas the real mean is probably not. Thus the variance (calculated from the square of the distance of each point to the mean) and hence standard deviation will probably be underestimated.

The quantity $(N-1)$ is known as the number of **degrees of freedom** of the sample. Since the concept of degrees of freedom is repeated throughout the rest of this book, it is important to describe what it means. In a sample of N observations each is free to have any value. However, if we have used the measurements to calculate the sample mean, this restricts the value the last point can have. Take a sample of two measurements, for instance. If the mean is 17 and the first measurement is $17 + 3 = 20$, the other measurement *must* have the value $17 - 3 = 14$. Thus, knowing the first measurement fixes the second, and there will only be one degree of freedom. In the same way, if you calculate the mean of any sample of size N, you restrict the value of the last measurement, so there will be only $(N - 1)$ degrees of freedom.

It can take time calculating the standard deviation by hand, but fortunately few people have to bother nowadays; estimates for the mean and standard deviation of the population can readily be found using computer statistics packages or even scientific calculators. All you need do is type in the data values and press the $\bar{x}$ button for the mean and the s or σ_{n-1} button for the population standard deviation. Do not use the σ_n button, since this is for equation (3.3) not equation (3.5).

Example 3.2

The masses (in tonnes) of a sample of 16 bull elephants from a single reserve in Africa were as follows.

4.5	5.2	4.9	4.3	4.6	4.8	4.6	4.9
4.5	5.0	4.8	4.6	4.6	4.7	4.5	4.7

Using a calculator, estimate the population mean and standard deviation.

Solution

The estimate for the population mean is 4.70 tonnes and the population standard deviation is 0.2251 tonne, rounded to 0.22 tonne to 2 decimal places. Note that both figures are given to one more degree of precision than the original data points because so many figures have been combined.

3.5 The variability of samples

It is relatively easy to calculate estimates of a population mean and standard deviation from a sample. Unfortunately, though, the estimate we calculated of the population mean $\bar{x}$ is unlikely to exactly equal the real mean of the population μ. In our elephant survey we might by chance have included more light elephants in our sample than one might expect, or more heavy ones. The estimate itself will be variable, just like the population. However, we can estimate how much it should vary and work out limits between which the mean is likely to be found.

3.5.1 The variability of samples from a known population

If we took an infinite number of samples from a population whose mean μ and standard deviation σ we knew, their means $\bar{x}$ would be normally

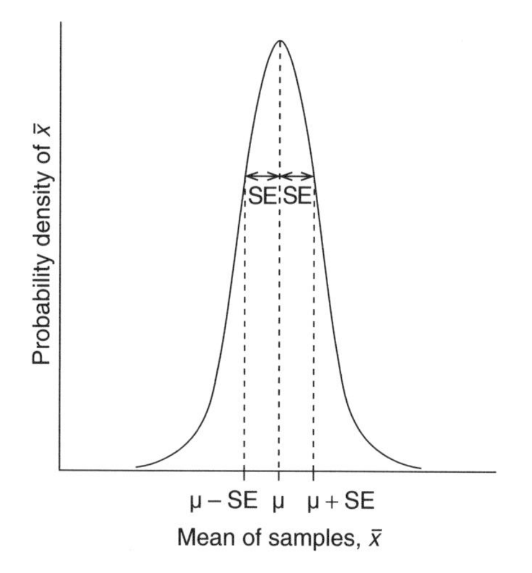

Figure 3.3 Distribution of sample means. The sample means $\bar{x}$ have a normal distribution with mean μ and standard error SE. The distribution is narrower than for single points (Figure 3.2) because, in a sample, high and low values tend to cancel each other out.

distributed just like the original measurements (Figure 3.3). However, the amount of spread of the means would be much narrower because high and low measurements would tend to cancel each other out in each sample, particularly in large samples. The **standard error** (SE) of the mean is a measure of how much the sample means would on average differ from the population mean. Just like standard deviation, standard error is the distance from the centre of the distribution to the inflexion point of the curve (Figure 3.3). It is given by the formula

$$\mathrm{SE} = \sigma/\sqrt{N} \qquad (3.6)$$

where σ is the standard deviation and N is the number of observations in the sample. Note that the bigger the sample size, the smaller the standard error. Just as we saw for standard deviation, 95% of the samples would have a mean within 1.96 times the standard error of the population, 99% within 2.58 times the standard error and 99.9% within 3.29 times the standard error. These limits are called 95%, 99% and 99.9% **confidence intervals**.

3.5.2 The variability of estimates of population means

There is a problem with calculating the variability of $\bar{x}$, like the value we calculated for the mass of elephants in Example 3.2. We do not know σ precisely, we only have our **estimate** s. However, we can still make an **estimate of the standard error**:

$$\overline{\mathrm{SE}} = s/\sqrt{N} \qquad (3.7)$$

Note that the larger the sample size, the smaller the value of $\overline{\mathrm{SE}}$.

LLYFRGELL COLEG MEN... ...ARY
BANGOR GWYNEDD ...

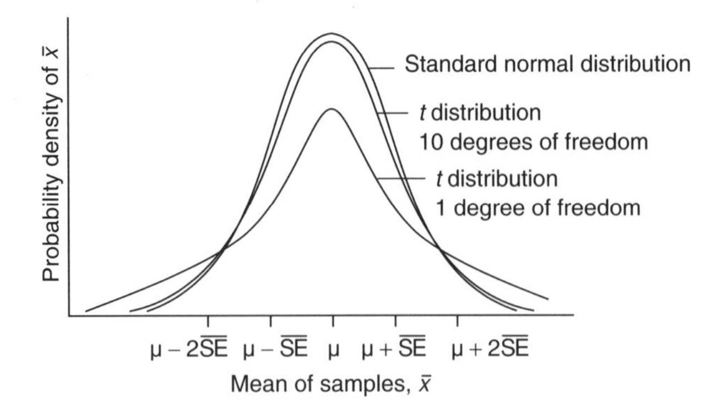

Figure 3.4 Normal distribution and *t* distribution. The distribution of sample means $\bar{x}$ relative to the estimate of the standard error $\overline{\mathrm{SE}}$ calculated from samples with 1, 10 and infinite degrees of freedom. With infinite degrees of freedom the distribution equals the normal distribution. However, it becomes more spread out as the sample size decreases (fewer degrees of freedom) because the estimate of standard error becomes less reliable.

3.6 Confidence limits for the population mean

Because standard error is only estimated, $\bar{x}$ will have a wider distribution relative to it than the normal distribution shown in Figure 3.3. In fact it will follow what is known as the ***t* distribution** (Figure 3.4). The exact shape of the *t* distribution depends on the number of degrees of freedom; it becomes progressively more similar to the normal distribution as the number of degrees of freedom increases (and hence as the estimate of standard deviation becomes more exact).

Knowing all this, it is fairly straightforward to calculate **confidence limits** for the population mean μ using the tabulated **critical values** of the *t* statistic given at the end of the book. The critical *t* value $t_{(N-1)}(5\%)$ is the number of standard errors $\overline{\mathrm{SE}}$ away from the estimate of population mean $\bar{x}$ within which the real population mean μ will be found 95 times out of 100. The 95% confidence limits define the 95% confidence interval, or 95% CI; this is expressed as follows:

$$95\%\ \mathrm{CI(mean)} = \bar{x} \pm (t_{(N-1)}(5\%) \times \overline{\mathrm{SE}}) \tag{3.8}$$

where $(N-1)$ is the number of degrees of freedom. It is most common to use a 95% confidence interval but it is also possible to calculate 99% and 99.9% confidence intervals for the mean by substituting the critical *t* values for 1% and 0.1% respectively into equation (3.8).

Note that the larger the sample size *N*, the narrower the confidence interval. This is because as *N* increases, not only will the standard error $\overline{\mathrm{SE}}$ be lower but so will the critical *t* values. Quadrupling the sample size reduces the distance between the upper and lower limits of the confidence interval by more than one-half.

Example 3.2

Our survey of bull elephants gave an estimate of mean mass of 4.70 tonnes and an estimate of standard deviation of 0.2251 tonne. We want to calculate 95% and 99% confidence limits for the mean mass.

Solution

The estimate of standard error is $\overline{SE} = 0.2251/\sqrt{16} = 0.0563$ tonne, which is rounded to 0.056 tonne to 3 decimal places. Notice that standard errors are usually given to one more decimal place than the mean or standard deviation.

To calculate the 95% confidence limits we must look in Table S1 (at the end of the book) for the critical value of t for $16 - 1 = 15$ degrees of freedom. In fact $t_{15}(5\%) = 2.131$. Therefore 95% confidence limits of the population mean are $4.70 \pm (2.131 \times 0.0563) = 4.70 \pm 0.12 = 4.58$ and 4.82 tonnes. So 95 times out of 100 the real population mean would be between 4.58 and 4.82 tonnes.

Similarly, $t_{15}(1\%) = 2.947$. Therefore 99% confidence limits of the population mean are $4.70 \pm (2.947 \times 0.0563) = 4.70 \pm 0.16 = 4.54$ and 4.86 tonnes. So 99 times out of 100 the real population mean would be between 4.54 and 4.86 tonnes.

3.7 The importance of descriptive statistics

We have seen that it is straightforward to calculate the mean, standard deviation, standard error of the mean and 95% confidence limits of a sample using a hand calculator. Together these sum up what you know about your sample and they are called **descriptive statistics**. Calculating them is the first and most important step in looking at the results of your surveys or experiments. You should work them out as soon as possible and try to see what they tell you.

3.8 Using computer packages

Nowadays you don't usually have to perform statistical calculations yourself. You can use one of the many computer-based statistical packages which are available, such as MINITAB, SPSS or EXCEL. You simply enter all your results straight into a spreadsheet in the computer package and let the computer take the strain. Using a package has two advantages: (1) the computer carries out the calculations more quickly; and (2) you can save the results for future analysis.

Most packages work in much the same way. You enter the results from different samples into their own separate columns. You can then run tests on the different columns from the command screen of the package.

Example 3.3

In order to calculate descriptive statistics for the data given in Example 3.2, the 16 values should be placed in rows 1 to 16 of the first column as follows:

		c1	c2	c3	c4	c5	c6	...
Rows	1	4.5						
	2	5.2						
	3	4.9						
	4	4.3						
	5	4.6						
	6	4.8						
	7	4.6						
	8	etc.						
	9	etc.						
	10	⋮						
	11							
	⋮							

It is then straightforward to investigate or to perform statistical tests on the data by running through menus or typing in commands. In MINITAB you can examine descriptive statistics by going to the Statistics menu, then choosing Basic Statistics and then Descriptives. Finally, select the column you want to examine. MINITAB will produce the following output:

	N	MEAN	MEDIAN	TRMEAN	STDEV	SE MEAN
bull	16	4.7000	4.6500	4.6929	0.2251	0.0563
	MIN	MAX	Q1	Q3		
bull	4.3000	5.2000	4.5250	4.8750		

It gives you the number, and estimates for the mean, standard deviation and standard error. Unfortunately, it also gives you some information you may not want, and it does not calculate the 95% confidence limits. The median is halfway between the eighth and ninth highest points; Q1 and Q3 are the points one-quarter and three-quarters of the way up the distribution. Their importance is explained in Chapter 7.

Note that the package gives some items with too much precision. Don't copy things from computer screens without thinking!

3.9 Presenting descriptive statistics

3.9.1 In text or tables

Once you have obtained your descriptive statistics, you need to express them in the correct way in your write-ups. There are two main ways of doing this.

The simplest is just to write them in your text or in tables as the mean followed by the standard deviation or the standard error in parentheses, e.g. $\bar{x}$ (s) or $\bar{x}$ ($\overline{\text{SE}}$). You must say whether you are giving the standard deviation or standard error and you must give the number of observations in your sample; this is so that the reader can calculate the other statistic. A 95% confidence interval can be given as $\bar{x} \pm (t_{(N-1)}(5\%) \times \overline{\text{SE}})$. For example, in our elephants example:

$$\text{Mean and standard deviation} = 4.70\ (0.22)\ \text{t}\ (n = 16)$$

$$\text{Mean and standard error} = 4.70\ (0.056)\ \text{t}\ (n = 16)$$

$$95\%\ \text{confidence interval} = 4.70 \pm 0.12\ \text{t}\ (n = 16)$$

3.9.2 Graphically

The other way to present data is on a point graph or a bar chart (Figure 3.5). The mean is the central point of the graph or the top of the bar.

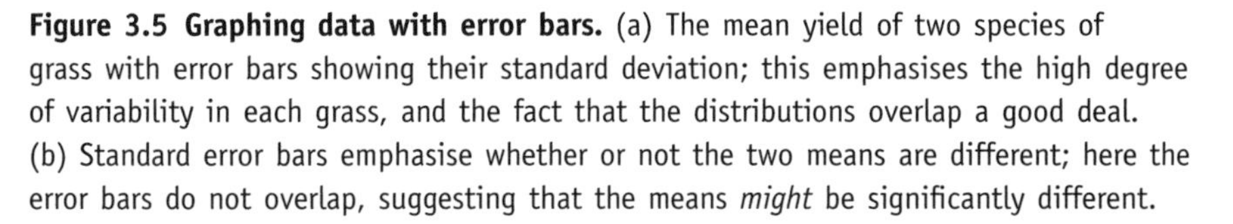

Figure 3.5 Graphing data with error bars. (a) The mean yield of two species of grass with error bars showing their standard deviation; this emphasises the high degree of variability in each grass, and the fact that the distributions overlap a good deal. (b) Standard error bars emphasise whether or not the two means are different; here the error bars do not overlap, suggesting that the means *might* be significantly different.

Error bars are then added. From the mean, bars are drawn both up and down a length equal to either the standard deviation or standard error. Finally, lines are drawn across the ends of the bars. Again you must say in the captions or legends which measure of error you are using.

The choice of standard deviation or standard error bars depends on what you want to emphasise about your results. If you want to show how much **variation** there is, you should choose standard deviation (Figure 3.5a). On the other hand, if you want to show how confident you can be of the mean, you should choose standard error (Figure 3.5b). In general, if two samples have overlapping standard error bars, they are unlikely to be statistically different (Chapter 4).

3.10 Self-assessment problems

Problem 3.1

In a population of women, heart rate is normally distributed with a mean of 75 and a standard deviation of 11. Between which limits will 95% of the women have their heart rates?

Problem 3.2

The masses (in grams) for a sample of 10 adult mice from a large laboratory population were measured. The following results were obtained:

5.6 5.2 6.1 5.4 6.3 5.7 5.6 6.0 5.5 5.7

Calculate estimates of the mean and standard deviation of the mass of the mice.

Problem 3.3

Nine measurements were taken of the pH of nine leaf cells. The results were as follows:

6.5 5.9 5.4 6.0 6.1 5.8 5.8 5.6 5.9

(a) Use the data to calculate estimates of the mean, standard deviation, and standard error of the mean. Use these estimates to calculate the 95% confidence interval for cell pH.

(b) Repeat the calculation assuming that you had only taken the first four measurements. How much wider is the 95% confidence interval?

Problem 3.4

The masses (in kilograms) of 25 newborn babies were as follows.

3.5	2.9	3.4	1.8	4.2	2.6	2.2	2.8	2.9	3.2	2.7	3.4	3.0
3.2	2.8	3.2	3.0	3.5	2.9	2.8	2.5	2.9	3.1	3.3	3.1	

Calculate the mean, standard deviation and standard error of the mean and present your results (a) in figures and (b) in the form of a bar chart with error bars showing standard deviation.

CHAPTER FOUR

Testing for differences

But 2-1 isn't a significant difference

4.1 Introduction

Even though any biological measurement is bound to be variable, by taking samples from a population, we can estimate the average value of the measurement, estimate its variability, and estimate the limits between which the average is likely to lie. This is very useful, but if we are going to carry out biological research, we might also want to answer specific questions about the things we are measuring. There are several questions we could ask:

- We might want to know if, and by how much, the average value of a measurement taken on a single population is different from an expected value. Is the birthweight of babies from a small town different from the national average?
- We might want to know if, on average, two measurements made on a single population are different from each other. Do patients have a different heart rate after taking beta blockers? Or is the pH of ponds different at dawn and dusk?
- We might want to know if experimentally treated organisms or cells are, on average, different from controls. Does shaking sunflowers alter their height compared with unshaken controls?
- We might want to know if two or more groups of organisms or cells are, on average, different from each other. Do different strains of bacteria have different growth rates?

This chapter describes how you can use statistical tests to help determine whether there are differences and how to work out their size.

4.2 Why we need statistical tests

4.2.1 The problem

You might imagine it would be easy to find out whether there are differences. You would just need to take measurements on your samples and compare the average values to see if they were different. However, there is a problem. Because of variation we can never be certain that the differences between our **sample means** reflect real differences in the **population means**. We might have got different means just by chance.

Suppose μ is the mean length of a population of rats. If we take measurements on a sample of rats, it is quite likely we could get a mean value $\bar{x}$ that is one standard error $\overline{SE}$ greater or smaller than μ. In fact the chances of getting a mean that different *or more* from μ is equal to the shaded area in Figure 4.1a. In contrast it is much less likely that we could get a value which is different by more than three standard errors from μ (Figure 4.1b). The probability is given by the tiny (though still real) area in the two tails of the distribution.

Therefore, if we take a sample and find that its mean is very different from an expected value, we can say that the mean of the population is very likely to be different from the expected value. However, we cannot be sure. Variation means we can never be sure that differences are real; and because

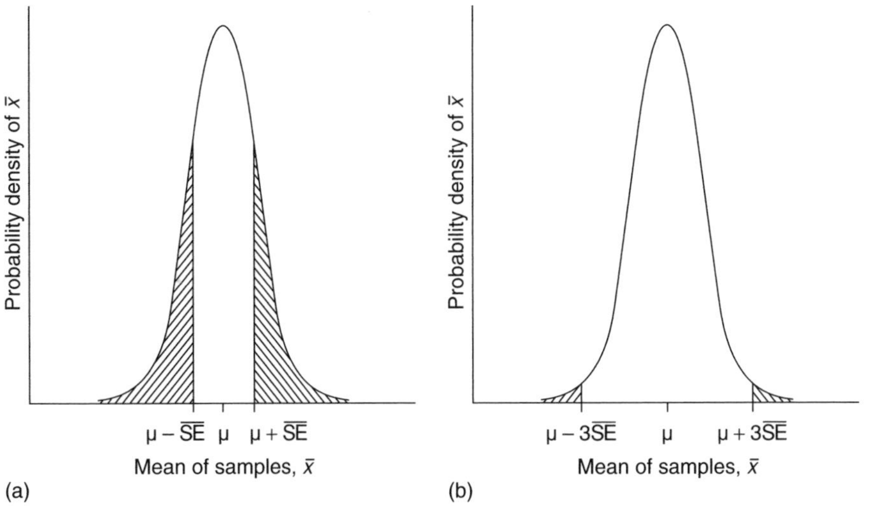

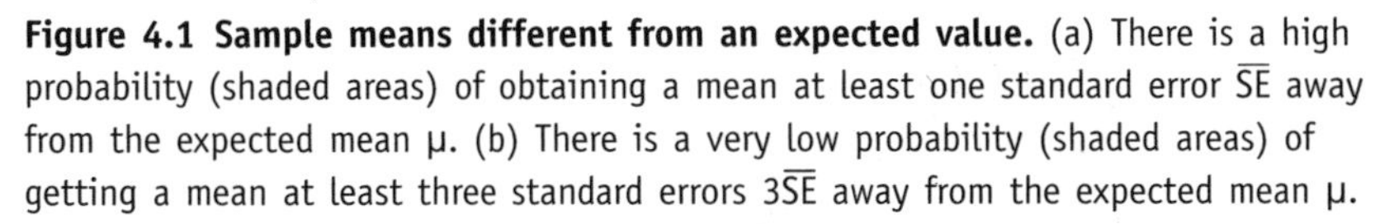

Figure 4.1 Sample means different from an expected value. (a) There is a high probability (shaded areas) of obtaining a mean at least one standard error $\overline{SE}$ away from the expected mean μ. (b) There is a very low probability (shaded areas) of getting a mean at least three standard errors $3\overline{SE}$ away from the expected mean μ.

LLYFRGELL COLEG MEN
BANGOR GWYNEDD

we can never be sure, we have to use statistical tests. These tests calculate just how likely it is that the differences are real, and how large any differences are likely to be.

4.2.2 The logic of statistics

All statistical tests use the same sort of counterintuitive logic, a logic which may seem difficult to follow. They all involve testing a hypothesis. This is done in four steps; a fifth step is also possible.

Step 1: Formulate a null hypothesis
The first stage is to assume the opposite of what you are testing. Here we are testing whether there is a difference, so we assume there is no difference. This is called the **null hypothesis**.

Step 2: Calculate the test statistic
The next stage is to examine the data values and calculate a test statistic from them. When testing for differences, the test statistic is usually a measure of how different the means are relative to the variability. The greater the difference in the means and the smaller the scatter in the data, the bigger the absolute value of the test statistic (i.e. the further away from zero it will be). The smaller the difference in the means and the greater the scatter, the smaller the absolute value of the test statistic.

Step 3: Calculate the significance probability
Next you must examine the test statistic and assess the probability of getting an absolute value that high or greater if the null hypothesis were true. The larger the absolute value of the test statistic (i.e. the further away from zero it is), hence the greater the distance between the means, the smaller the probability. The smaller the absolute value of the test statistic, the larger the probability.

Step 4: Decide whether to reject the null hypothesis

- If the significance probability is below a **critical value**, you must reject the null hypothesis and conclude that there is a **significant difference**. Usually in biology one rejects the null hypothesis if the significance probability is less than 1 in 20. This probability is often written as the decimal 0.05, or as 5%. This criterion for rejecting the null hypothesis is therefore known as the 5% significance level.
- If the significance probability is greater than 5%, you have no evidence to reject the null hypothesis. But this does not mean you have evidence to support it.

Statisticians have taken a lot of the hard work out of calculating the significance probability and deciding whether to reject the null hypothesis by preparing tables of critical values for test statistics. Three of these tables are given at the end of the book. All you need do is consult the appropriate table for the critical value. Statistical tables often come in two different

Table 4.1 Four conventional ways to present the results of significance tests.[a]

Probability is less than 1 in 20	*Probability is less than 1 in 100*	*Probability is less than 1 in 1000*
5% level	1% level	0.1% level
$P < 0.05$	$P < 0.01$	$P < 0.001$
$0.01 < P < 0.05$	$0.001 < P < 0.01$	$P < 0.001$
*	**	***

[a] The second is the most commonly used.

versions: one-tailed and two-tailed. In this book we use the **two-tailed tests**, by far the more common in biology; they test whether there are differences from expected values but not which sign they are. With our rats, therefore, we would be testing whether they had a different length but not whether they were longer or shorter than expected. The criterion for rejecting the null hypothesis in the two-tailed test is when the total area in the two tails of the distribution (Figure 4.1) is less than 5%, so each tail must have an area of less than 2.5%.

Sometimes you may find the probability P falls below critical levels of 1 in 100 or 1 in 1000. If this is true, you can reject the null hypothesis at the 1% or 0.1% levels respectively. There are several ways in which significance levels may be presented in scientific papers and reports. Four of the commonest ways are shown in Table 4.1. The second way is the most common.

Step 5: Calculate confidence limits

Whether or not there is a **significant difference**, you can calculate **confidence limits** to give a set of plausible values for the differences of the means. Calculating 95% confidence limits for the difference of means is just as straightforward as calculating 95% confidence limits for the means themselves (Section 3.6).

This may sound rather abstract, so let's look at some examples. We begin with the simplest of all tests for differences, the one-sample t test.

4.3 The one-sample *t* test

4.3.1 Purpose

To test whether the sample mean of one measurement taken on a single population is different from an expected value E.

4.3.2 Rationale

You work out how many standard errors the sample mean is away from the expected value. The further away it is, the less probable it is that the real mean is the expected value.

4.3.3 Carrying out the test

Step 1: Formulate the null hypothesis
The null hypothesis is that the mean of the population *is not* different from the expected value.

Step 2: Calculate the test statistic
The test statistic t is the number of standard errors the sample mean is away from the expected value. It can be found using a calculator or using MINITAB.

Using a calculator

$$t = \frac{\text{Sample mean} - \text{Expected value}}{\text{Standard error of mean}} = \frac{\bar{x} - E}{\text{SE}} \tag{4.1}$$

Note that t could be positive or negative. It is the difference from zero which matters.

Using MINITAB
Statistical packages such as MINITAB can readily work out t as well as other important elements in the test. Simply put the data into a column, go into the Statistics menu, then into Basic Statistics and choose 1-Sample t. Finally, select the column you want to test and the test value you want to compare it with.

Step 3: Calculate the significance probability
You must calculate the probability P that the absolute value of t, written $|t|$, would be this high or greater if the null hypothesis were true.

Using a calculator
You must compare your value of $|t|$ with the critical value of the t statistic for $(N-1)$ degrees of freedom and at the 5% level ($t_{(N-1)}(5\%)$). This is given in Table S1 (at the end of the book).

Using MINITAB
MINITAB will directly work out the probability P. (Note that the bigger the value of $|t|$, the smaller the value of P.)

Step 4: Decide whether to reject the null hypothesis

Using a calculator

- If $|t|$ is greater than the critical value, you must reject the null hypothesis. Therefore you can say that the mean is significantly different from the expected value.
- If $|t|$ is less than the critical value, you have no evidence to reject the null hypothesis. Therefore you can say that the mean is not significantly different from the expected value.

Using MINITAB

- If $P < 0.05$ you must reject the null hypothesis. Therefore you can say that the mean is significantly different from the expected value.
- If $P > 0.05$ you have no evidence to reject the null hypothesis. Therefore you can say that the mean is not significantly different from the expected value.

Step 5: Calculate confidence limits
The 95% confidence limits for the difference are given by the equation

$$95\%\ \mathrm{CI(difference)} = \bar{x} - E \pm (t_{(N-1)}(5\%) \times \overline{\mathrm{SE}}) \qquad (4.2)$$

Example 4.1

Do the bull elephants we first met in Example 3.2 have a different mean mass from the mean value for the entire population of African elephants of 4.50 tonnes?

Solution

Formulate the null hypothesis
The null hypothesis is that the mean weight of bull elephants *is* 4.50 tonnes.

Calculate the test statistic

Using a calculator
The mean weight $\bar{x}$ of the sample of bull elephants is 4.70 tonnes with an estimate of the standard error $\overline{\mathrm{SE}}$ of 0.0563 tonnes. Therefore $t = (4.70 - 4.50)/0.0563 = 3.55$. The mean is 3.55 standard errors away from the expected value.

Using MINITAB
MINITAB produces the following output:

```
TEST OF MU = 4.5000 VS MU N.E. 4.5000
          N      MEAN     STDEV    SE MEAN      T    P VALUE
bull     16    4.7000    0.2251     0.0563   3.55     0.0029
```

Calculate the significance probability

Using a calculator
Looking up in the t distribution for $16 - 1 = 15$ degrees of freedom, the critical value that $|t|$ must exceed for the probability to drop below the 5% level is 2.131.

Using MINITAB
The MINITAB printout shows that the probability P of $|t|$ being this high or greater is $P = 0.0029$.

Decide whether to reject the null hypothesis

Using a calculator

$$|t| = 3.55 > 2.131$$

Using MINITAB

$$P = 0.0029 < 0.05$$

Therefore we must reject the null hypothesis. We can say that bull elephants have a weight significantly different from 4.50 tonnes; in fact they are heavier.

Calculate confidence limits

The difference between the actual and expected mean = 4.70 − 4.50 = 0.20, the standard error is 0.0563, and the critical t value for 15 degrees of freedom is 2.131. Therefore

$$95\% \text{ CI(difference)} = 0.20 \pm (2.131 \times 0.0563) = 0.08 \text{ to } 0.32$$

Bull elephants are 95% likely to be between 0.08 and 0.32 tonnes heavier than 4.5 tonnes.

4.4 The paired *t* test

4.4.1 Purpose

To test whether the means of **two** measurements made on a **single** identifiable population are different from each other.

4.4.2 Rationale

This test has two stages. You first calculate the difference d between the two measurements you have made on each item. You then use a one-sample t test to determine whether the mean difference $\bar{d}$ is different from zero.

4.4.3 Carrying out the test

Step 1: Formulate the null hypothesis

The null hypothesis is that the mean difference $\bar{d}$ *is not* different from zero.

Step 2: Calculate the test statistic

The test statistic t is the number of standard errors the difference is away from zero. It can be calculated using a calculator or using MINITAB.

Using a calculator

$$t = \frac{\text{Mean difference}}{\text{Standard error of difference}} = \frac{\bar{d}}{\text{SE}_d} \quad (4.3)$$

Using MINITAB
Statistical packages such as MINITAB can readily work out t as well as other important elements in the test. Simply put the data side by side into two columns and subtract one from the other to form a 'difference' column. Finally, from the Statistics menu choose Basic Statistics and then 1-Sample t. This test is used to determine whether the mean of this column is different from zero.

Step 3: Calculate the significance probability
You must calculate the probability P that the absolute value of the test statistic would be equal to or greater than t if the null hypothesis were true.

Using a calculator
You must compare your value of $|t|$ with the critical value of the t statistic for $(N-1)$ degrees of freedom and at the 5% level ($t_{(N-1)}(5\%)$). This is given in Table S1 (at the end of the book).

Using MINITAB
MINITAB will directly work out the probability P. (Note that the bigger the value of $|t|$, the smaller the value of P.)

Step 4: Decide whether to reject the null hypothesis

Using a calculator

- If $|t|$ is greater than the critical value, you must reject the null hypothesis. Therefore you can say that the mean difference is significantly different from zero.
- If $|t|$ is less than the critical value, you have no evidence to reject the null hypothesis. Therefore you can say that the mean difference is not significantly different from zero.

Using MINITAB

- If $P < 0.05$ you must reject the null hypothesis. Therefore you can say that the mean difference is significantly different from zero.
- If $P > 0.05$ you have no evidence to reject the null hypothesis. Therefore you can say that the mean is not significantly different from zero.

Step 5: Calculate confidence limits
The 95% confidence limits for the mean difference are given by the equation

$$95\%\ \mathrm{CI(difference)} = \bar{d} \pm (t_{(N-1)}(5\%) \times \overline{\mathrm{SE}}_d) \tag{4.4}$$

Example 4.2

Two series of measurements were made of the pH of nine ponds: at dawn and at dusk. The results are shown below. Do the ponds have a different pH at these times?

Pond	*Dawn pH*	*Dusk pH*	*Difference*
1	4.84	4.91	0.07
2	5.26	5.62	0.36
3	5.03	5.19	0.16
4	5.67	5.89	0.22
5	5.15	5.44	0.29
6	5.54	5.49	−0.05
7	6.01	6.12	0.11
8	5.32	5.61	0.29
9	5.44	5.70	0.26
$\bar{x}$	5.362	5.552	0.190
s	0.352	0.358	0.129
$\overline{\text{SE}}$	0.1174	0.1194	0.0431

Carrying out descriptive statistics shows that the mean difference $\bar{d} = 0.19$ and the standard error of the difference $\overline{\text{SE}}_d = 0.043$.

Solution

Formulate the null hypothesis

The null hypothesis is that the mean of the differences in the pH *is* 0, i.e. the ponds have the same pH at dawn and dusk.

Calculate the test statistic

Using a calculator

We have $t = 0.190/0.0431 = 4.40$. The difference is 4.40 standard errors away from zero.

Using MINITAB

MINITAB produces the following output:

```
TEST OF MU = 0.0000 VS MU N.E. 0.0000
                N      MEAN      STDEV     SE MEAN      T     P VALUE
dusk-dawn       9     0.1900    0.1294     0.0431     4.40    0.0023
```

Calculate the significance probability

Using a calculator

Looking up in the t distribution for $9 - 1 = 8$ degrees of freedom, the critical value of t for the 5% level is 2.306.

Using MINITAB

MINITAB shows that the probability P of $|t|$ being this high or greater is $P = 0.0023$.

Decide whether to reject the null hypothesis

Using a calculator

$$|t| = 4.40 > 2.306$$

Using MINITAB

$$P = 0.0023 < 0.05$$

Therefore we must reject the null hypothesis. We can say that the mean difference between dawn and dusk is significantly different from 0. In other words, the pH of ponds is significantly different at dusk from at dawn; in fact it's higher.

Calculate confidence limits

The 95% confidence interval is $0.19 \pm (2.306 \times 0.043) = 0.09$ to 0.29. It is 95% likely that the pH at dusk will be between 0.09 and 0.29 higher than the pH at dawn.

4.5 The two-sample t test

4.5.1 Purpose

To test whether the means of a single measurement made on two populations are different from each other.

4.5.2 Rationale

This test is rather more complex than the previous two because you have to decide the probability of overlap between the distributions of *two* sample means (Figure 4.2). To do this you have to calculate t by comparing the difference in the means of the two populations with an estimate of the **standard error of the difference** between the two populations. The test makes the assumption that the variances of the two populations are the same.

The two-sample t test also makes an important assumption about the measurements: it assumes the two sets of measurements are **independent** of each other. This would not be true of the data on the ponds we examined in Example 4.2, because each measurement has a pair, a measurement from the same pond at a different time of day. Therefore it is not valid to carry out a two-sample t test on this data.

4.5.3 Carrying out the test

Step 1: Formulate the null hypothesis

The null hypothesis is that the mean of the differences *is not* different from zero. In other words, the two groups have the same mean.

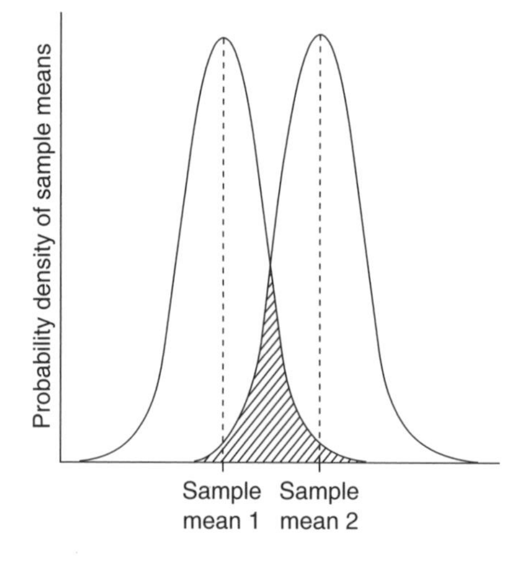

Figure 4.2 Overlapping populations. The estimated probability distributions of two overlapping populations worked out from the results of samples.

Step 2: Calculate the test statistic
The test statistic t is given by the formula

$$t = \frac{\text{Mean difference}}{\text{Standard error of difference}} = \frac{\bar{x}_A - \bar{x}_B}{\overline{\text{SE}}_d} \tag{4.5}$$

In this case it is much more complex to calculate the standard error of the difference $\overline{\text{SE}}_d$ because this would involve comparing each member of the first population with each member of the second. But $\overline{\text{SE}}_d$ can be estimated if we assume that the variance of the two populations is the same. It is given by the equation

$$\overline{\text{SE}}_d = \sqrt{(\overline{\text{SE}}_A)^2 + (\overline{\text{SE}}_B)^2} \tag{4.6}$$

where $\overline{\text{SE}}_A$ and $\overline{\text{SE}}_B$ are the standard errors of the two populations. If the populations are of similar size, $\overline{\text{SE}}_d$ will be about $1\frac{1}{2}$ times as big as either population standard error.

Fortunately, computer statistics packages can perform all the calculations almost instantaneously. To perform a test in MINITAB, put the data into two columns, go into the Statistics menu, then Basic Statistics and choose 2-Sample t. Finally select the columns you want to test. You can perform the test with or without making the assumption that the variances are the same.

Step 3: Calculate the significance probability
You must calculate the probability P that the absolute value of the test statistic would be equal to or greater than t if the null hypothesis were true. MINITAB will directly work this out. There are $N_A + N_B - 2$ degrees of freedom, where N_A and N_B are the sample sizes of groups A and B.

Step 4: Decide whether to reject the null hypothesis

- If $P < 0.05$ you must reject the null hypothesis. Therefore you can say that the sample means are significantly different from each other.
- If $P > 0.05$ you have no evidence to reject the null hypothesis. Therefore you can say that the two sample means are not significantly different from each other.

Step 5: Calculate confidence limits

The 95% confidence intervals for the mean difference are given by the equation

$$95\%\ \mathrm{CI(difference)} = \bar{x}_A - \bar{x}_B \pm (t_{N_A+N_B-2}(5\%) \times \overline{\mathrm{SE}}_d) \qquad (4.7)$$

But MINITAB also calculates these limits directly.

Example 4.3

The following data were obtained by weighing 16 cow elephants as well as the 16 bull elephants we have already weighed. We will test whether bull elephants have a different mean mass from cow elephants.

Masses of bull elephants (tonnes)

4.5	5.2	4.9	4.3	4.6	4.8	4.6	4.9
4.5	5.0	4.8	4.6	4.6	4.7	4.5	4.7

Masses of cow elephants (tonnes)

4.3	4.6	4.5	4.4	4.7	4.1	4.5	4.4
4.2	4.3	4.5	4.4	4.5	4.4	4.3	4.3

Solution

Carrying out descriptive statistics yields the following results:

Bull elephants: mean = 4.70, $s = 0.22$, $\overline{\mathrm{SE}} = 0.056$

Cow elephants: mean = 4.40, $s = 0.15$, $\overline{\mathrm{SE}} = 0.038$

It looks like bulls are heavier, but are they significantly heavier?

Formulate the null hypothesis

The null hypothesis is that the mean of the differences in weight is 0, i.e. bull and cows have the same mean weight.

LLYFRGELL COLEG MENAI LIBRARY
BANGOR GWYNEDD LL57 2TP

Calculate the test statistic

MINITAB comes up with the following output:

```
TWOSAMPLE T FOR bulls VS cows
              N          MEAN          STDEV          SE MEAN
bulls         16         4.700         0.225          0.056
cows          16         4.400         0.151          0.038

95 PCT CI FOR MU bulls - MU cows: (0.161, 0.439)

TTEST  MU  bulls = MU  cows  (Pooled  St  Dev = 0.191):  T = 4.43
P = 0.0001 DF = 30
```

MINITAB has calculated that $t = 4.43$. The means are 4.43 standard errors of the difference away from each other.

Calculate the significance probability

MINITAB has calculated the probability P directly as $P = 0.0001$.

Decide whether to reject the null hypothesis

We have $P = 0.0001 < 0.05$. Therefore we must reject the null hypothesis. We can say that bull elephants have a significantly different mean weight from cow elephants; in fact they are heavier.

Calculate confidence limits

MINITAB has also calculated that the 95% confidence interval for the difference is 0.161 to 0.439. The weight difference between bull and cow elephants has a 95% probability of being between 0.161 and 0.439 tonnes.

4.6 ANOVA: comparing many groups

4.6.1 Why t tests are unsuitable

If you want to compare the means of more that two groups, you might think that you could simply compare each group with all the others using two-sample t tests. However, there are two good reasons why you should not do this. First, there is the problem of convenience. As the number of groups you are comparing goes up, the number of tests you must carry out rises rapidly, from 3 tests when comparing 3 groups to 45 tests for 10 groups.

Number of groups	3	4	5	6	7	8	9	10
Number of t tests	3	6	10	15	21	28	36	45

But there is a second, more important problem. We reject a null hypothesis with 95% confidence, not 100% confidence. This means that in 1 in 20 tests we will falsely assume there is a significant difference between groups when none really exists. If we carry out a lot of tests, the chances of making such an error go up rapidly, so if we carry out 45 tests there is about a 90% chance we will find significant effects even if none exist.

For these reasons you must use a more complex statistical test to determine whether there is a difference between many groups; it is called analysis of variance, usually shortened to **ANOVA**.

4.6.2 The rationale behind one-way ANOVA

One-way ANOVA works in a different manner from *t* tests. Rather than examine the difference between the means directly, ANOVA looks at the **variability** of the data. Let's examine a simple example in which the means of the weights of just two small samples of fish are compared (Figure 4.3a). The overall variability is the sum of the squares of the distances from each point to the overall mean (Figure 4.3b); here it's $3^2 + 2^2 + 1^2 + 3^2 + 2^2 + 1^2 = 28$. But this can be split into two parts. First, there is the between-group variability, which is due to the differences between the group means. This is the sum of the squares of the distances of each point's group mean from the overall mean (Figure 4.3c); here it's $(6 \times 2^2) = 24$. Second, there is the within-group variability, which is due to the scatter within each group. This is the sum of the squares of the distance from each point to its group mean (Figure 4.3d); here it's $(4 \times 1^2) + (2 \times 0^2) = 4$.

ANOVA compares the between-group variability and the within-group variability. To show how this helps, let's look at two contrasting situations. In Figure 4.4a the two means are far apart and there is little scatter within each group; the between-group variability will clearly be much larger than the within-group variability. In Figure 4.4b the means are close together and there is much scatter within each group; the between-group variability will be lower than the within-group variability.

4.6.3 Carrying out one-way ANOVA

The actual workings of ANOVA tests are actually a little more complex, but they involve the same four basic steps as the *t* tests we have already carried out.

Step 1: Formulate the null hypothesis

The null hypothesis is that the groups have the same mean. In this case the hypothesis is that the two groups of fish have the same mean weight.

Step 2: Calculate the test statistic

The test statistic in ANOVA tests is the *F* statistic. Calculating *F* is quite a complex process; it involves producing a table like the one shown in Example 4.4.

(a) The first stage is to calculate the variabilities due to each factor to produce the so-called sums of squares (SS).
(b) We cannot compare sums of squares, because they are the result of adding up different numbers of points. The next stage is therefore to calculate

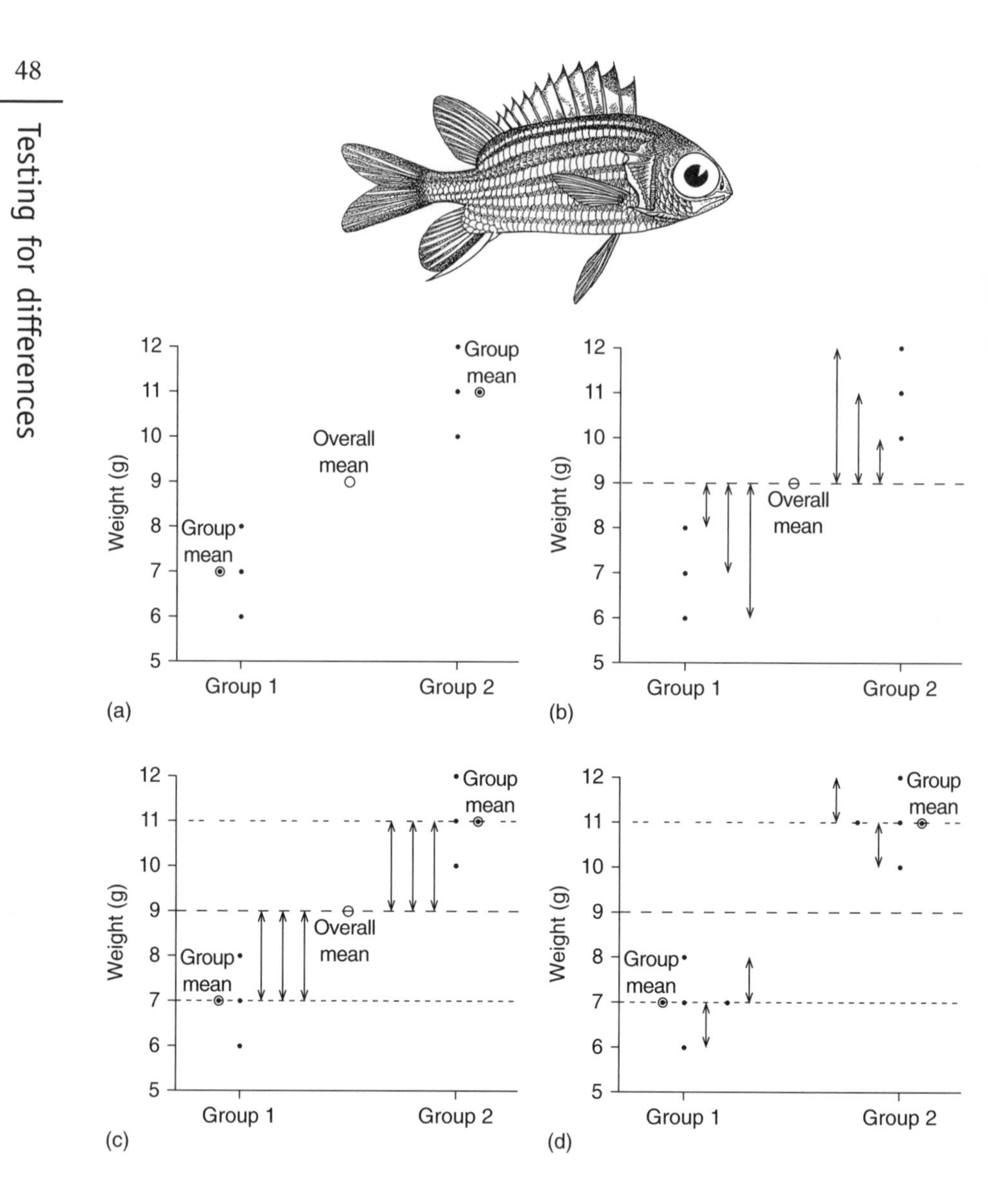

Figure 4.3 The rationale behind ANOVA: hypothetical weights for two samples of fish. (a) Calculate the overall mean and the group means. (b) The total variability is the sum of the squares of the distances of each point from the overall mean; this can be broken down into between-group variability and within-group variability. (c) The between-group variability is the sum of the squares of the distances from each point's group mean to the overall mean. (d) The within-group variability is the sum of the squares of the distances from each point to its group mean.

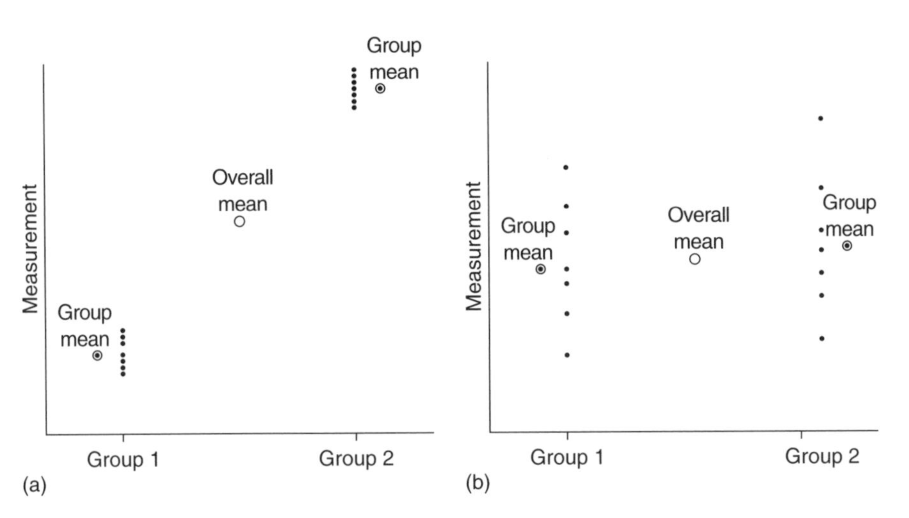

Figure 4.4 Two contrasting situations. (a) Most of the variability is caused by the group means being far apart. (b) Most of the variability is caused by differences within the groups.

the actual **variance** or **mean squares** (MS) due to each factor. This is calculated by dividing each sum of squares by the correct number of **degrees of freedom**.

(i) If there are n groups, the between-group degrees of freedom $DF_B = n - 1$.
(ii) If there are N items in total and r items in each group, there will be $r - 1$ degrees of freedom in each group, hence $n(r - 1)$ in total. The within-group degrees of freedom, $DF_W = N - n$.
(iii) If there are N items in total, the total number of degrees of freedom $DF_T = N - 1$.

(c) The last stage is to calculate the test statistic F. This is the ratio of the between-group mean square MS_B to the within-group mean square MS_W.

$$F = MS_B/MS_W \qquad (4.9)$$

The larger the value of F, the more likely it is that the means are significantly different.

These calculations tend to be extremely lengthy. Therefore it is best to carry out ANOVA tests on statistical packages such as MINITAB. Simply enter the data from each sample into different columns, go into the Statistics menu, then ANOVA and choose Oneway (separate columns). Finally, select the columns you want to test.

Step 3: Calculate the significance probability

Tables do exist to determine how large F values have to be for different degrees of freedom, but I recommend that you only carry out ANOVA tests using computer statistics packages. They will automatically calculate the significance probability P.

Step 4: Decide whether to reject the null hypothesis

- If $P < 0.05$ you must reject the null hypothesis.
- If $P > 0.05$ you have no evidence to reject the null hypothesis.

Step 5: Calculate confidence limits and test which groups are different
The problem with the ANOVA test is that though it tells us whether there are differences between groups, it doesn't tell us how big the differences are or even which groups are different. If your ANOVA test is significant, however, you can carry out one of several **post hoc tests**, such as the **Tukey test** or the **least significant difference test**, to answer these questions. Unfortunately, there is not enough space to describe them here.

Example 4.4: Fish data in Figure 4.3

Formulate the null hypothesis

The null hypothesis is that the two groups of fish have the same mean weight.

Calculate the test statistic

MINITAB will come up with the following results:

```
ANALYSIS OF VARIANCE
SOURCE    DF     SS      MS       F       P
BETWEEN    1   24.00   24.00   24.00   0.008
WITHIN     4    4.00    1.00
TOTAL      5   28.00
                               INDIVIDUAL 95 PCT CI'S FOR MEAN
                               BASED ON POOLED STDEV
LEVEL      N    MEAN   STDEV   ---------+---------+---------+-------
Group 1    3   2.000   1.000    (-------*-------)
Group 2    3   6.000   1.000                      (-------*-------)
                               ---------+---------+---------+-------
POOLED STDEV =1.000                    2.0       4.0       6.0
```

Hence we can see that $F = 24$.

Calculate the significance probability

Here $P = 0.008$. Note that the error bars are not overlapping; this suggests there is likely to be a significant difference.

Decide whether to reject the null hypothesis

In our example $P = 0.008 < 0.05$, so we must reject the null hypothesis. We can say that the two groups of fish have significantly different mean weights.

4.6.4 Problems with names

ANOVA is hard enough to understand anyway. Unfortunately, things are made more difficult because, for historical reasons, there are two synonyms for *between* and *within*:

between = treatment = factor

within = error = residual

You must be able to recognise all of them. Then you can cope with any statistics book or any lecturer!

4.7 Further uses of ANOVA

ANOVA is so counterintuitive and complex that many students are inclined to forget all about it. However, this would be a mistake because it can be used for many other things than just comparing the means of a number of groups. It allows you to carry out and analyse a whole new range of experiments in which you can look at the effect of two or more factors at once:

- You might want to examine the effect on corn yield of adding different amounts of nitrogen and phosphorus.
- You might want to examine the effect on yield of adding different amounts of nitrogen to more than one wheat variety.

In the first case, let's say you grow wheat at two different levels of nitrogen and at two different levels of phosphorus. As long as you grow at all possible combinations of nitrogen and phosphorus levels (so there are $2 \times 2 = 4$ combinations in total), and you have the same number of replicates for each combination, you can analyse such an experiment using what is called **two-way ANOVA**. The yields ($t\ ha^{-1}$) from just such an experiment are tabulated here.

No nitrate or phosphate Mean = 1.88, $s = 0.32$, $\overline{SE} = 0.105$	1.4	1.8	2.1	2.4	1.7	1.9	1.5	2.0	2.1
Added nitrate only Mean = 2.80, $s = 0.24$, $\overline{SE} = 0.082$	2.4	2.7	3.1	2.9	2.8	3.0	2.6	3.1	2.6
Added phosphate only Mean = 3.44, $s = 0.40$, $\overline{SE} = 0.132$	3.5	3.2	3.7	2.8	4.0	3.2	3.9	3.6	3.1
Added nitrate and phosphate Mean = 6.88, $s = 0.61$, $\overline{SE} = 0.203$	7.5	6.4	8.1	6.3	7.2	6.8	6.4	6.7	6.5

You can carry out two-way ANOVA using statistical packages such as MINITAB and get results in a table like the ones we have already seen. The results for our yield data are as follows.

LLYFRGELL COLEG MENAI LIBRARY
BANGOR GWYNEDD LL57 2TP

ANALYSIS OF VARIANCE FOR YIELD

Source	DF	SS	MS	F	P
Nitrogen	1	42.684	42.684	248.65	0.000
Phosphorus	1	71.684	71.684	417.58	0.000
Interaction	1	14.188	14.188	82.65	0.000
Error	32	5.493	0.172		
Total	35	134.050			

Just like the one-way ANOVA we have already looked at, two-way ANOVA partitions the variability and variance. However, there will be not two possible causes of variability but four: the effect of nitrogen; the effect of phosphorus; the **interaction** between the effects of nitrogen and phosphorus; and finally, variation within the groups (here called error).

These possibilities can be used to produce three *F* ratios, which can answer three questions in just one experiment:

1. Does nitrogen significantly affect yield?
2. Does phosphorus significantly affect yield?
3. Do nitrogen and phosphorus interact?

Here the *F* ratios are all high and the significance probabilities are low. In this case all three terms are clearly significant. So what does this mean?

1. It is clear from looking at the descriptive statistics that adding nitrogen increases yield (by 0.92 t ha^{-1}).
2. Looking at the descriptive statistics, it is also clear that adding phosphorus increases yield (by 1.56 t ha^{-1}).
3. Looking at the descriptive statistics, you can also tell that adding both nitrogen and phosphorus together increases yield much more than just the sum of the effects of nitrogen and phosphorus alone (by 5.0 t ha^{-1} rather than by just $0.92 + 1.56 = 2.48$ t ha^{-1}). In this case, the significant interaction term shows that the two fertilisers potentiate each other's effects and so act synergistically.

The interaction term would also have been significant if the two fertilisers inhibited each other's effect, causing a reduced yield when added together.

4.8 Self-assessment problems

Problem 4.1

The scores (in percent) of 25 students in a statistics test were as follows.

58	65	62	73	70	42	56	53	59	56	60	64	63
78	90	31	65	58	59	21	49	51	58	62	56	

Calculate the mean, standard deviation and standard error of the mean for these scores. The mean mark of students in finals exams is supposed to be 58%. Perform a one-sample t test to determine whether these students did significantly differently from expected.

Problem 4.2

The masses (in grams) of 16 randomly chosen tomatoes grown in a commercial glasshouse were as follows.

32	56	43	48	39	61	29	45
53	38	42	47	52	44	36	41

Other growers have found that the mean mass of this sort of tomato is 50 g. Perform a one-sample t test to determine whether the mean mass of tomatoes from this glasshouse is different from the expected mass. Give the 95% confidence intervals for the mean mass.

Problem 4.3

Students were tested on their ability to predict how moving bodies behave, both before and after attending a course on Newtonian physics. Their marks are tabulated here. Did attending the course have a significant effect on their test scores, and if so by how much?.

	Before	*After*
Martha	45	42
Denise	56	50
Betty	32	19
Amanda	76	78
Eunice	65	63
Ivy	52	43
Pamela	60	62
Ethel	87	90
Letitia	49	38
Patricia	59	53

Problem 4.4

The pH of cactus cells was measured at dawn and at dusk using microprobes. The following results were obtained.

Dawn	5.3	5.6	5.2	7.1	4.2	4.9	5.4	5.7	6.3	5.5	5.7	5.6
Dusk	6.7	6.4	7.3	6.2	5.2	5.9	6.2	6.5	7.6	6.4	6.5	

(a) Using a statistical package such as MINITAB, carry out a two-sample *t* test to determine if there is any significant difference in pH between the cells at these times.
(b) The cactus was identifiable and two sets of measurements were carried out on it. So why can't you analyse this experiment using the paired *t* test?

Problem 4.5

An experiment was carried out to investigate the effect of mechanical support on the yield of wheat plants. The masses of seed (in grams) produced by 20 control plants and 20 plants whose stems had been supported throughout their life were as follows.

Control	9.6	10.8	7.6	12.0	14.1	9.5	10.1	11.4	9.1	8.8
	9.2	10.3	10.8	8.3	12.6	11.1	10.4	9.4	11.9	8.6
Supported	10.3	13.2	9.9	10.3	8.1	12.1	7.9	12.4	10.8	9.7
	9.1	8.8	10.7	8.5	7.2	9.7	10.1	11.6	9.9	11.0

Using a statistical package such as MINITAB, carry out a two-sample *t* test to determine whether support has a significant effect on yield.

Problem 4.6

The effect of three different antibiotics on the growth of a bacterium was examined by adding them to Petri dishes, which were then inoculated with the bacteria. The diameter of the colonies (in millimetres) was then measured after three days. A control where no antibiotics were added was also included. The following results were obtained.

Control	4.7	5.3	5.9	4.6	4.9	5.0	5.3	4.2
	5.7	5.3	4.6	5.8	4.7	4.9		
Antibiotic A	4.5	5.6	5.4	4.9	4.8	4.6	5.1	5.3
	5.3	5.8	5.1	6.0	4.4	5.3		
Antibiotic B	4.7	5.2	5.4	4.4	6.1	4.8	5.3	5.5
	4.7	5.2						
Antibiotic C	4.3	5.7	5.3	5.6	4.5	4.9	5.1	5.3
	4.7	6.3	4.8	4.9	5.2	5.4	4.8	5.0

Carry out a one-way ANOVA test on a statistical package to determine whether there is any evidence that the antibiotic treatments affected the growth of the bacteria.

Problem 4.7

Interpret the following ANOVA table. How many groups were being compared? What was the total number of observations? And was there a significant difference between the groups?

	DF	SS	MS	F	P
Factor	4	654	164	1.71	0.35
Residual	25	2386	95		
Total	29	3040			

Problem 4.8

In a second experiment, two different varieties of wheat, Widgeon and Hereward, were grown at three different levels of nitrogen. The following results were obtained.

Widgeon	*Nitrates added (kg m^{-2})*		
	0	*1*	*2*
Yield (t ha^{-1})	4.7	6.4	7.8
	5.3	7.5	7.4
	5.1	6.9	8.3
	6.0	8.1	6.9
	6.5	5.9	6.5
	4.8	7.6	7.2
	5.6	7.1	6.3
	5.8	6.4	7.9
	5.4	8.6	7.7

Hereward	*Nitrates added (kg m^{-2})*		
	0	*1*	*2*
Yield (t ha^{-1})	1.3	6.1	10.8
	2.2	7.2	9.8
	2.1	7.4	11.4
	3.3	8.6	10.6
	1.8	5.7	12.2
	2.4	7.2	9.6
	2.6	6.7	11.1
	2.7	6.9	10.4
	3.1	8.4	10.9

Carrying out two-way ANOVA in MINITAB yielded the following results:

ANALYSIS OF VARIANCE FOR YIELD

Source	DF	SS	MS	F	P
nitrogen	2	240.911	120.456	208.65	0.000
variety	1	0.145	0.145	0.25	0.618
nitrogen*variety	2	95.189	47.595	82.44	0.000
Error	48	27.711	0.577		
Total	53	363.957			

(a) Examine the results of the experiment and the ANOVA table. Which of the three possible effects, variety, nitrogen and interaction, are significant?

(b) Examine the original data to work out what these results mean in real terms.

CHAPTER FIVE

Finding associations

An experiment to test whether the more really is the merrier

5.1 Introduction

We saw in the last chapter that one can use a paired t test to determine whether two measurements taken on a single group of organisms are different. For instance, one can test whether students have a different heart rate after drinking coffee compared with before. But we may instead want to know if and how the two measurements are **associated**. Do the students who have a higher heart rate before drinking coffee have a higher heart rate afterwards as well? Or we might ask other questions. How are the lengths of snakes related to their age? How is the wing area of birds related to their weight? Or how is the blood pressure of stroke patients related to their heart rate?

This chapter has three sections. First, it shows how to examine data to see whether variables are associated. Second, it describes some of the ways in which biological variables can be related. Finally, despite the inevitable variability, it shows how you can use statistical tests to work out whether there is a real association between the variables, and how to determine what it is.

5.2 Examining data for associations

The first thing you should do if you feel that two variables might be associated is to draw a **scatter plot** of one against the other. This will allow you to see at a glance what is going on. For instance, it is clear from Figure 5.1 that as the age of eggs increases, their mass decreases. But it is important

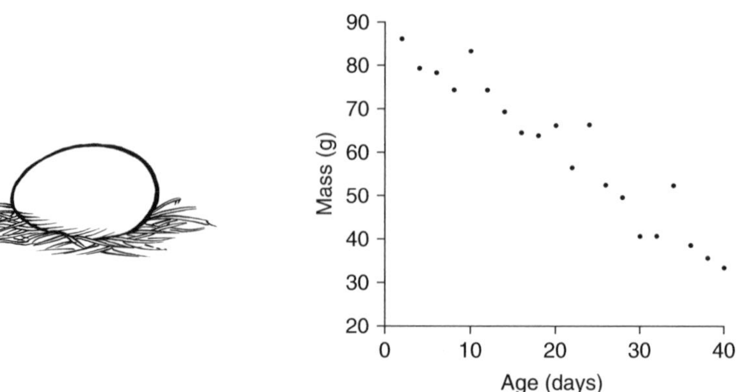

Figure 5.1 The relationship between the age of eggs and their mass. Note that the dependent variable, mass, is plotted along the vertical axis.

to make sure you plot the graph the correct way round. This depends on how the variables affect each other. One of the variables is called the independent variable; the other variable is called the dependent variable. The independent variable affects, or may affect, the dependent variable but is not itself affected. Plot the **independent variable** along the horizontal axis, often called the *x*-axis. Plot the **dependent variable** along the vertical axis, often called the *y*-axis. You would then say you were plotting the dependent variable against the independent variable. In Figure 5.1, age is the independent variable and mass is the dependent variable. This is because age can affect an egg's mass, but mass can't affect an egg's age.

Things are not always so clear-cut. It is virtually impossible to tell whether blood pressure would affect heart rate or vice versa. They are probably both affected by a third variable – artery stiffness. In this case, it does not really matter which way round you plot the data.

5.3 Examining graphs

Once you have plotted your graph, you should examine it for associations. There are several main ways in which variables can be related:

- There may be no relationship: points are scattered all over the graph paper (Figure 5.2a).
- There may be a positive association (Figure 5.2b): the dependent variable increases as the independent variable increases.
- There may be a negative association (Figure 5.2c): the dependent variable decreases as the independent variable increases.
- There may be a more complex relationship: Figure 5.2d shows a relationship in which the dependent variable rises and falls as the independent variable increases.

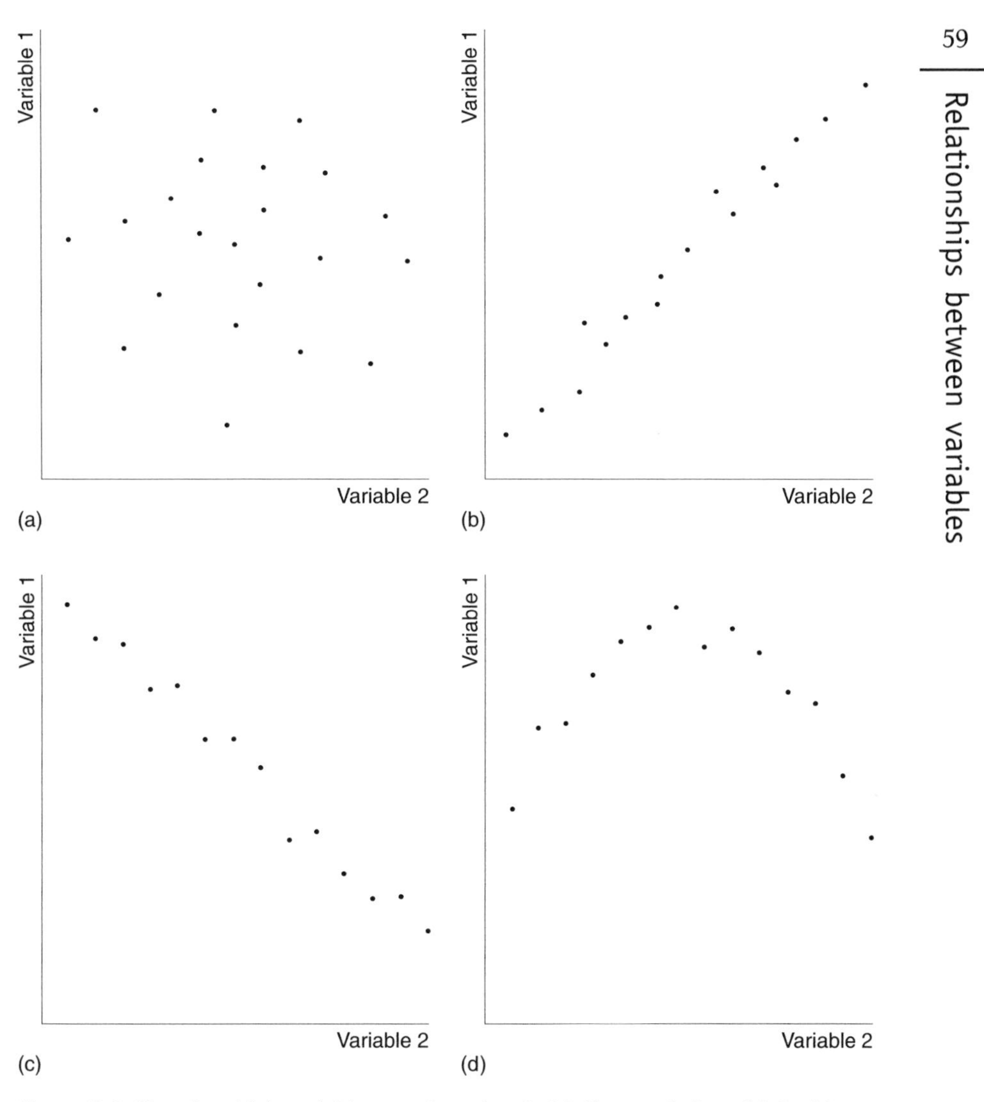

Figure 5.2 Ways in which variables can be related. (a) No association. (b) Positive association. (c) Negative association. (d) A complex curvilinear association.

5.4 Relationships between variables

There are an infinite number of ways in which two variables can be related. However, most are complex. Perhaps the simplest relationships to describe are linear relationships such as the one shown in Figure 5.3. In these cases, the dependent variable y is related to the independent variable x by the general equation

$$y = a + bx \tag{5.1}$$

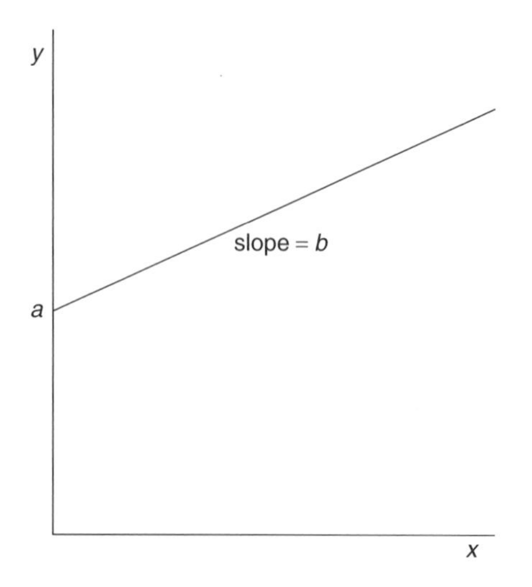

Figure 5.3 A straight line relationship. The straight line $y = a + bx$ has y-intercept a and slope b.

where b is the **slope** of the line and a is the **intercept**. The intercept is the value of y where the line crosses the y-axis.

Linear relationships are important because they are by far the easiest to analyse statistically. When biologists test whether two variables are related, they are usually testing whether they are linearly related. Fortunately, linear relationships between variables are surprisingly common in biology. Many other common relationships between variables can also be converted into linear relationships by **transforming** the data using logarithms.

5.4.1 Scaling and power relationships

If you examine organisms of different size, many of their characteristics scale according to **power relationships**. If an organism changes in size by a given ratio, some characteristic will increase or decrease by, the square, cube or some other power of that ratio. For instance, the mass of unicellular algae would be expected to rise with the cube of their diameter; and the metabolic rate of mammals rises with mass to the power of around 0.75. Other physical processes are also related in this way. The lift produced by a bird's wings should rise with the square of the flight speed.

In these sorts of relationships, the dependent variable y is related to the independent variable x by the general equation

$$y = ax^b \tag{5.2}$$

Looking at the curves produced by this sort of relationship (Figure 5.4a), it is very difficult to determine the values of a and b. However, it is possible, by using some clever mathematical tricks, to produce a straight line graph from which a and b can be easily calculated. The first thing to do is to take logarithms of both sides of the equation. We have $y = ax^b$, so

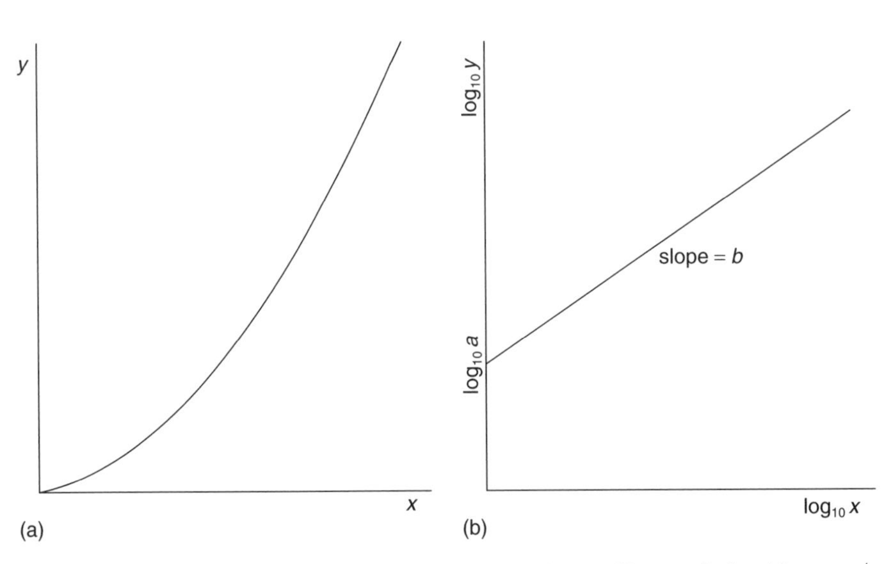

Figure 5.4 How to describe a power relationship. The curvilinear relationship $y = ax^b$ (a) can be converted to a straight line (b) by taking logarithms ($\log_{10}$) of both x and y. The graph has y-intercept $\log_{10} a$ and slope b.

$$\log_{10} y = \log_{10}(ax^b) \tag{5.3}$$

Now logarithms have two important properties:

$$\log_{10}(c \times d) = \log_{10} c + \log_{10} d \tag{5.4}$$

$$\log_{10}(c^d) = d \times \log_{10} c \tag{5.5}$$

Using these properties we can rearrange the equation to show that

$$\log_{10} y = \log_{10} a + b \log_{10} x \tag{5.6}$$

Therefore plotting $\log_{10} y$ against $\log_{10} x$ (Figure 5.4b) will produce a straight line with slope b and intercept $\log_{10} a$.

5.4.2 Exponential growth and decay

Other biological phenomena have an **exponential relationship** with time. In these cases, when a given period of time elapses, some characteristic increases or decreases by a certain ratio. For instance, bacterial colonies demonstrate exponential growth, doubling in number every few hours. In contrast, radioactivity shows exponential decay, halving over a given period. Other physical processes are also related in this way. Rates of reaction, indeed the metabolic rates of whole organisms, increase exponentially with temperature.

In these sorts of relationship the dependent variable y can be related to the independent variable x by the general equation

$$y = ae^{bx} \tag{5.7}$$

where e is the base of natural logarithms ($e = 2.718$), which we met in Section 2.9.2. Looking at the curve produced by this sort of relationship (Figure 5.5a),

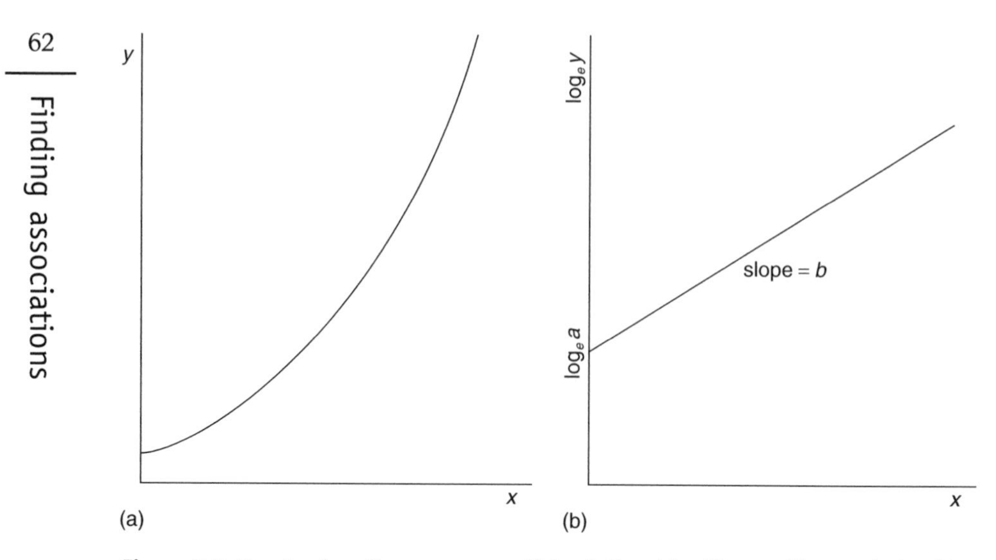

Figure 5.5 How to describe an exponential relationship. The curvilinear relationship $y = ae^{bx}$ (a) can be converted to a straight line (b) by taking natural logarithms ($\log_e$) of y. The graph has y-intercept $\log_e a$ and slope b.

it is very difficult to determine the values of a and b, just as it was with power relationships. However, we can again use some clever mathematical tricks to produce a straight line graph. As before, the first thing to do is to take logarithms of both sides of the equation. Therefore

$$\log_e y = \log_e(ae^{bx}) \tag{5.8}$$

and rearranging

$$\log_e y = \log_e a + bx \tag{5.9}$$

Therefore plotting $\log_e y$ against x (Figure 5.5b) will produce a straight line with slope b and y-intercept $\log_e a$.

5.5 Statistical tests for associations

The points on your plots will never exactly follow a straight line, or indeed any exact mathematical function, because of the variability which is inherent in biology. There will always be some scatter away from a line. The difficulty in determining whether two measurements are really associated is that when you were taking a sample you might have chosen points which followed a straight line even if there was no association in the population. If there appears only to be a slight association and if there are only a few points, this is quite likely. In contrast it is very unlikely that you would choose large numbers of points all along a straight line just by chance if there was no real relationship. Therefore you have to carry out statistical tests to work out the probability you could get your apparent association by chance. If there is an association, you can then work out what it is. This involves testing hypotheses and finding confidence limits. Testing for associations uses the same counterintuitive statistical logic we met in the last chapter. There are two main tests for association: correlation and regression.

5.6 Correlation

5.6.1 Purpose

To test whether two sets of measurements taken on a single population are linearly associated.

5.6.2 The rationale behind correlation

Correlation analysis examines the extent to which two sets of measurements show positive or negative association. The basic idea is that if there is positive association, all points will either be above and to the right or below and to the left of the distribution's centre (Figure 5.6a). If there is negative

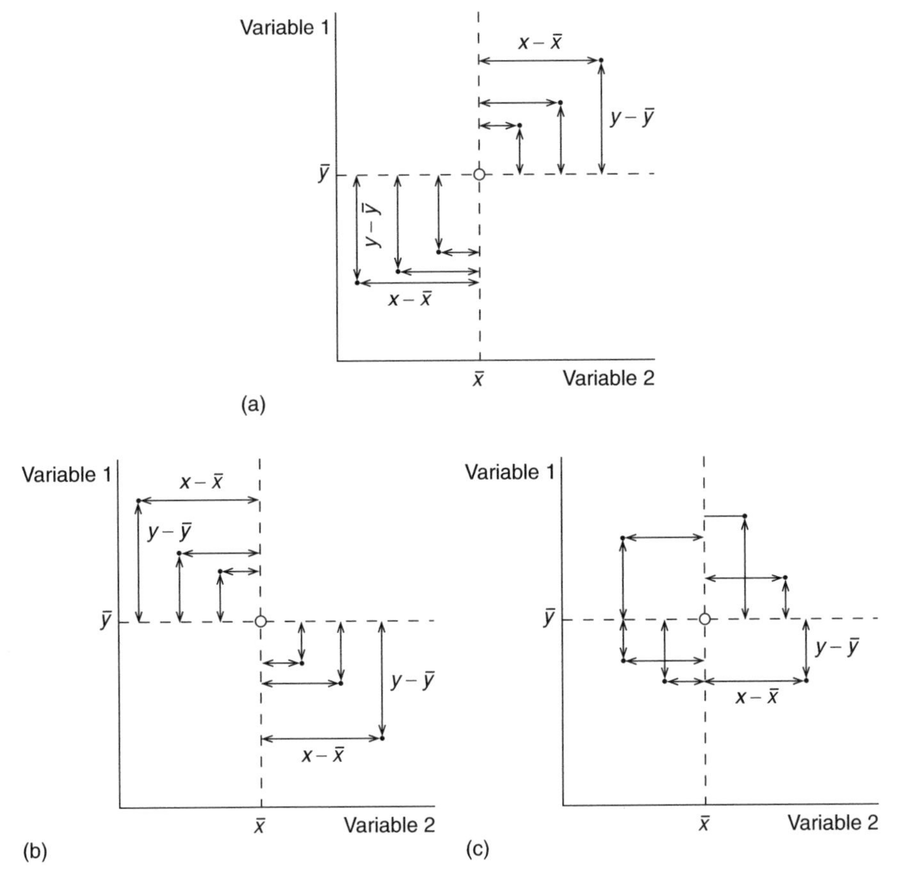

Figure 5.6 Correlation. (a) Positive correlation: $\Sigma(x-\bar{x})(y-\bar{y})$ is large and positive. (b) Negative correlation: $\Sigma(x-\bar{x})(y-\bar{y})$ is large and negative. (c) No correlation: $\Sigma(x-\bar{x})(y-\bar{y})$ is small.

association, all points will be above and to the left or below and to the right of the distribution's centre (Figure 5.6b). If there is no association, the points will be scattered on all sides (Figure 5.6c).

5.6.3 Carrying out the test

Correlation analysis follows all four of the usual steps of statistical tests.

Step 1: Formulate the null hypothesis
In correlation, the null hypothesis is that there is no association between the two measurements, i.e. they show random scatter.

Step 2: Calculate the test statistic
The test statistic for correlation is the correlation coefficient r. It is calculated in three stages.

Stage 1
Calculate the means of the two sets of measurements, $\bar{x}$ and $\bar{y}$.

Stage 2
For each point, calculate the product of its x and y distances from the mean $(x - \bar{x})(y - \bar{y})$. Note that if both x and y are greater than the mean, this figure will be positive because both $(x - \bar{x})$ and $(y - \bar{y})$ will be positive. It will also be positive if both x and y are smaller than the mean, because both $(x - \bar{x})$ and $(y - \bar{y})$ will be negative and their product will be positive. However, if one is larger than the mean and the other smaller, the product will be negative.

These points are added together to give

$$\text{Sum} = \Sigma(x - \bar{x})(y - \bar{y})$$

- If there is positive association (Figure 5.6a), with points all either above and to the right or below and to the left of the overall mean, the sum will be large and positive.
- If there is negative association (Figure 5.6b), with points all either above and to the left or below and to the right of the overall mean, the sum will be large and negative.
- If there is no association (Figure 5.6c), points will be on all sides of the overall mean, and the positive and negative numbers will cancel each other out. The sum will therefore be small.

Stage 3
Scale the sum obtained in stage 2 by dividing it by the product of the variation within each of the measurements. The correlation coefficient r is therefore given by the formula

$$r = \frac{\Sigma(x - \bar{x})(y - \bar{y})}{[\Sigma(x - \bar{x})^2 \Sigma(y - \bar{y})^2]^{1/2}} \qquad (5.10)$$

The correlation coefficient can vary from −1 (perfect negative correlation) through 0 (no correlation) up to a value of 1 (perfect positive correlation). The calculation is somewhat involved, so it is recommended that you carry out correlation analysis on a computer package such as MINITAB. Simply put your data into two columns; from the Statistics menu choose Basic Statistics and then Correlation. Finally, select the two columns and run the test.

Step 3: Calculate the significance probability
You must calculate the probability P that the absolute value of the correlation coefficient $|r|$ would be equal to or greater than a critical value if the null hypothesis were true. The larger the value of $|r|$ and the larger the sample, the less likely this will be. Critical values of $|r|$ required for P to fall below 0.05, and hence for the association to be significant, are given for a range of degrees of freedom in Table S2.

Since MINITAB does not calculate P, you must look up in Table S2 (at the end of the book) the critical value of r for $(N-2)$ degrees of freedom, where N is the number of pairs of observations.

Step 4: Decide whether to reject the null hypothesis
- If $|r|$ is greater than the critical value, you must reject the null hypothesis. You can say that the two variables show significant correlation.
- If $|r|$ is less than the critical value, you cannot reject the null hypothesis. There is no evidence of a linear association between the two variables.

Example 5.1

In an investigation of the cardiovascular health of elderly patients, the heart rate and blood pressure of 30 patients were taken. The following results were obtained. Is there any association between the variables?

Patient	*Heart rate* (min^{-1})	*Blood pressure (mm Hg)*
1	67	179
2	75	197
3	63	175
4	89	209
5	53	164
6	76	180
7	98	212
8	75	187
9	71	189
10	65	176
11	69	167
12	74	186
13	80	198

Patient	*Heart rate* (min^{-1})	*Blood pressure (mm Hg)*
14	58	170
15	76	187
16	68	175
17	64	169
18	76	190
19	79	176
20	72	168
21	60	158
22	67	160
23	63	167
24	90	221
25	50	149
26	73	180
27	64	168
28	68	162
29	65	168
30	70	157

Solution

Formulate the null hypothesis

The null hypothesis is that blood pressure and heart rate are not associated.

Calculate the test statistic

MINITAB comes up with the following output:

```
Correlation of Heart Rate and Blood Pressure = 0.866
```

Calculate the significance probability

Looking up r in Table S2 (at the end of the book), the critical value that r must exceed at $30 - 2 = 28$ degrees of freedom is about 0.381.

Decide whether to reject the null hypothesis

We have $|r| = 0.866 > 0.381$. Therefore we must reject the null hypothesis. We can say that heart rate and blood pressure show a significant positive association.

5.6.4 Uses of the correlation coefficient

Correlation is a useful technique since it tells you whether two measurements are associated, and it can be used even if neither of the variables is independent of the other. However, the results of correlation analysis need to be treated with caution for three reasons:

- Correlation only finds linear associations between measurements, so a non-significant correlation does not prove there is no association between the variables.
- A significant correlation does not imply a **causal relationship** between the two measurements.
- The size of the correlation coefficient does not reflect the slope of the relationship between the two measurements, it just reflects how close the association is. If you want to determine the nature of the linear relationship between two sets of measurements, you need to carry out regression analysis. However, this is only valid if one of the variables is obviously independent of the other and so is plotted along the x-axis of your graph.

5.7 Regression

5.7.1 Purpose

To quantify the linear relationship between two sets of measurements taken on a single population.

5.7.2 Rationale

Regression analysis finds an estimate of the line of best fit $y = \bar{a} + \bar{b}x$ through the scattered points on your graph. If you measure the vertical distance of each point from the regression line (Figure 5.7a), the line of best fit is the one which minimises the sum of the squares of the distances.

The estimate of the slope $\bar{b}$ is actually worked out in a similar way to the correlation coefficient, using the formula

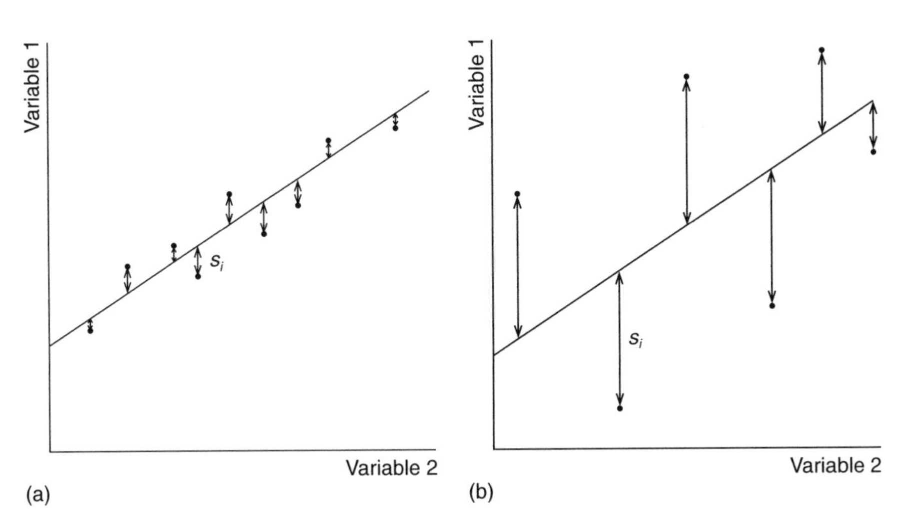

Figure 5.7 Regression. The line of best fit minimises the variability Σs_i^2 from the line. (a) Significant regression: Σs_i^2 is low. (b) Non-significant regression: Σs_i^2 is high.

$$\bar{b} = \frac{\Sigma(x - \bar{x})(y - \bar{y})}{\Sigma(x - \bar{x})^2} \tag{5.11}$$

Since the line of best fit always passes through the means of x and y, $\bar{x}$ and $\bar{y}$, the estimate of the constant $\bar{a}$ can then be found by substituting them into the equation to give

$$\bar{a} = \bar{y} - \bar{b}\bar{x} \tag{5.12}$$

This is all very well, but data with very different degrees of scatter, such as those shown in Figure 5.7, can have identical regression lines. In Figure 5.7a there is clearly a linear relationship. However, in Figure 5.7b there may actually be no relationship between the variables; you might have chosen a sample that suggests there is a relationship just by chance.

In order to test whether there is really a relationship, therefore, you would have to carry out one or more statistical tests. You could do this yourself, but the calculations needed are a bit long and complex, so it is much better now to use a computer package. MINITAB not only calculates the regression equation but also performs two statistical tests and gives you the information you need to carry out a whole range of other tests:

- It works out the **standard deviation** of $\bar{a}$ and $\bar{b}$ and uses them to carry out two separate ***t* tests** to determine whether they are significantly different from zero. The data can also be used to calculate 95% confidence intervals for a and b.
- It carries out an **ANOVA** test, which essentially compares the amount of variation explained by the regression line with that due to the scatter of the points away from the regression line. This tells you whether there is a significant slope, e.g. if $\bar{b}$ is significantly different from zero.
- It also tells you the percentage of the total variation which the regression line explains. This r^2 value is equal to the square of the correlation coefficient.

5.7.3 Carrying out the test

Step 1: Formulate the null hypotheses
The null hypotheses MINITAB tests are (i) y does not depend on x, so the true slope of the line is zero ($b = 0$); and (ii) the intercept is zero ($a = 0$).

Step 2: Calculate the test statistics
In MINITAB first enter your data into two columns. Then choose Regression from the Statistics menu. Put the dependent variable into the response box and the independent variable into the predictor box. Run the test.

Step 3: Calculate the significance probability
The next thing to read off is the probability P that absolute values of the test statistics would be equal to or greater than t and F respectively if the null hypotheses were true.

Step 4: Decide whether to reject the null hypothesis

- If $P < 0.05$ you should reject the null hypothesis. Therefore you can say that the slope (or intercept) is significantly different from zero.
- If $P > 0.05$ you have no evidence to reject the null hypothesis. Therefore you can say that the slope (or intercept) is not significantly different from zero.

Step 5: Calculate confidence limits

To work out how different the slope is from zero, we must work out 95% confidence limits for the slope b. These are calculated in just the same way as we calculated 95% confidence limits for means in Chapter 4:

$$95\%\ \mathrm{CI(slope)} = \bar{b} \pm (t_{(N-2)}(5\%) \times \overline{\mathrm{SD}}\ \text{of slope}) \tag{5.13}$$

Here $t_{(N-2)}$ is the critical t value for $(N-2)$ degrees of freedom, where N is the number of data points. Confidence limits can also be calculated for the intercept.

Example 5.2

In a survey to investigate the way in which chicken eggs lose weight after they are laid, one egg was collected newly laid every 2 days. Each egg was put into an incubator, and after 40 days all 20 eggs were weighed. The results are tabulated here and plotted in Figure 5.1. Carry out a regression analysis to determine whether age significantly affects egg weight. If there is a relationship, determine what it is.

Age (days)	2	4	6	8	10	12	14	16	18	20	22
Mass (g)	87	80	79	75	84	75	70	65	64	67	57

Age (days)	24	26	28	30	32	34	36	38	40
Mass (g)	67	53	50	41	41	53	39	36	34

Solution

Formulate the null hypothesis

The null hypothesis is that age has no effect on egg weight. In other words, the slope of the regression line is zero.

Calculate the test statistic

Carrying out the test in MINITAB yields the following output:

```
THE REGRESSION EQUATION IS
Mass = 89.4 - 1.36 Age
Predictor          Coef          Stdev          t ratio         P
Constant         89.437          2.279            39.24      0.000
Age             -1.36128        0.09514          -14.31      0.000
s = 4.907            R-sq = 91.9%              R-sq(adj) = 91.5%
```

ANALYSIS OF VARIANCE

SOURCE	DF	SS	MS	F	P
Regression	1	4929.2	4929.2	204.74	0.000
Error	18	433.4	24.1		
Total	19	5362.6			

The slope of the regression equation is −1.36, which appears to be well below zero. But is this difference significant? Looking at the probabilities that the slope is zero, we have to look at the t ratio for age and the F value for the analysis of variance table. Here $t = -14.31$ and $F = 204.74$ (note that $F = t^2$).

Calculate the significance probability

Looking at the t and F probabilities, both equal 0.000.

Decide whether to accept or reject the null hypothesis

We have $P = 0.000 < 0.05$. Therefore we must reject the null hypothesis. We can say that age has a significant effect on egg weight; in fact older eggs are lighter.

Calculate confidence limits

We need to look up the critical values for t for $20 - 2 = 18$ degrees of freedom in Table S1. Using equation (5.13)

$$95\%\ \text{CI(slope)} = -1.36 \pm (t_{18}(5\%) \times 0.09514)$$

$$95\%\ \text{CI(slope)} = -1.36 \pm (2.101 \times 0.09514)$$

$$= -1.56 \text{ to } -1.16$$

5.7.4 Other tests on regression data

The t tests worked out by the computer investigate whether the slope and constant are different from zero. The value of t is simply given by the expression

$$t = \frac{\text{Observed value} - 0}{\text{Standard deviation}} \tag{5.14}$$

However, it is also possible from MINITAB output to carry out a whole range of t tests to determine whether the slope or constant are different from any expected value. Then t is simply given by the expression

$$t = \frac{\text{Observed value} - \text{Expected value}}{\text{Standard deviation}} \tag{5.15}$$

and you can carry out the t test for $N - 2$ degrees of freedom just as the computer did to determine whether the slope or constant were different from zero.

Example 5.3

From the egg weight data in Example 5.2, we want to determine whether the initial egg weight was significantly different from 90 g, which is the mean figure for the general population. In other words, we must test whether the intercept (or constant as MINITAB calls it) is different from 90.

Solution

Formulate the null hypothesis
The null hypothesis is that the constant is equal to 90.

Calculate the test statistic
The necessary data can be extracted from the MINITAB output in Example 5.2 (see p. 69). This shows that the estimate of the intercept = 89.437 and its standard deviation = 2.279. Therefore if the expected value = 90, the test statistic is

$$t = \frac{89.437 - 90}{2.279}$$

$$= -0.247$$

Calculate the significance probability
Here $|t|$ must be compared with the critical value for $20 - 2 = 18$ degrees of freedom. This is 2.101.

Decide whether to reject the null hypothesis
We have $|t| = 0.247 < 2.101$. Hence there is no evidence to reject the null hypothesis. We can say that initial egg mass is not significantly different from 90 g.

5.7.4 Validity

You must be careful to use regression appropriately; there are many cases where it is not valid:

- Regression is not valid for data in which there is no independent variable. For example, you should not regress heart rate against blood pressure, because each factor could affect the other.
- All your measurements must be independent. Therefore you should not use regression to analyse repeated measures, such as the height of a single plant at different times.

5.8 Self-assessment problems

Problem 5.1

Which way round would you plot the following data?

(a) Cell number of an embryo and time since fertilisation
(b) Pecking order of hens and of their chicks
(c) Height and body weight of women
(d) Length and breadth of limpets

Problem 5.2

(a) The logarithms of the wing area A of birds and their body length L are found to be related by the straight line relationship $\log_{10} A = 0.3 + 2.36 \log_{10} L$. What is the relationship between A and L?
(b) The natural logarithm of the numbers of cells N in a bacterial colony is related to time T by the equation $\log_e N = 2.3 + 0.1T$. What is the relationship between N and T?

Problem 5.3

A study of the density of stomata in vine leaves of different areas came up with the following results. Calculate the correlation coefficient r between these two variables and determine whether this is a significant correlation. What can you say about the relationship between leaf area and stomatal density?

Leaf area (mm^2)	45	56	69	32	18	38	48	26	60	51
Stomatal density (mm^{-2})	36	28	27	39	56	37	32	45	24	31

Problem 5.4

In a survey to investigate why bones become more brittle in older women, the density of bone material was measured in 24 post-menopausal women of contrasting ages. Bone density is given as a percentage of the average density in young women.

Age (years)	43	49	56	58	61	63	64	66	68	70	72	73
Relative bone density	108	85	92	90	84	83	73	79	80	76	69	71
Age (years)	74	74	76	78	80	83	85	87	89	92	95	98
Relative bone density	65	64	67	58	50	61	59	53	43	52	49	42

(a) Plot the data.
(b) Using a statistics package, carry out a regression analysis to determine the relationship between age and bone density. Does bone density change significantly with age?
(c) Calculate the expected bone density of women of age 70.

Problem 5.5

In an experiment to examine the ability of the polychaete worm *Nereis diversicolor* to withstand zinc pollution, worms were grown in solutions containing different concentrations of zinc and their internal zinc concentration was measured. The following results were obtained.

$\log_{10}$ [Zn]$_{water}$	1.96	2.27	2.46	2.65	2.86	2.92	3.01	3.24	3.37	3.49
$\log_{10}$ [Zn]$_{worm}$	2.18	2.23	2.22	2.27	2.25	2.30	2.31	2.34	2.36	2.35

(a) Plot the data.
(b) Using a statistics package, carry out a regression analysis to determine how zinc in the solution affects the concentration within the worm. If *Nereis* did not actively control its level of zinc, the concentrations inside and outside would be equal and the slope of the regression line would be 1. Work out the t value which compares a slope of 1 with the slope of the line you obtained, hence determine whether *Nereis* actively controls its zinc level.

Problem 5.6

A study of the effect of seeding rate on the yield of wheat gave the following results.

Seeding rate (m^{-2})	50	80	100	150	200	300	400	500	600	800
Yield (tonnes)	2.5	3.9	4.7	5.3	5.6	5.9	5.4	5.2	4.6	3.2

(a) Plot a graph of yield against seeding rate.
(b) Using a statistics package, carry out regression analysis to determine whether there is a linear relationship between seeding rate and yield.
(c) What can you say about the relationship between seeding rate and yield?

Problem 5.7

An investigation was carried out into the scaling of heads in worker army ants. Body length and jaw width were measured in 20 workers of contrasting size. The following results were obtained.

Length (mm)	3.2	3.6	4.2	4.3	4.6	5.0	5.2	5.3	5.5	5.5
Jaw width (mm)	0.23	0.29	0.32	0.38	0.45	0.44	0.55	0.43	0.60	0.58
Length (mm)	5.7	6.2	6.6	6.9	7.4	7.6	8.5	9.2	9.7	9.9
Jaw width (mm)	0.62	0.73	0.74	0.88	0.83	0.93	1.03	1.15	1.09	1.25

(a) Convert the data to logarithms ($\log_{10}$) and plot them.

(b) Using a statistics package, carry out regression analysis to investigate the relationship between $\log_{10}$(length) and $\log_{10}$(jaw width). Convert this equation back to real numbers.

(c) Perform a t test on the slope of the regression line to determine whether it is different from 1. Do the jaws of the ants remain the same relative size, or do they get significantly smaller or larger in bigger ants?

CHAPTER SIX

Dealing with categorical data

Among professors there are significantly more goats than expected

6.1 Introduction

Often in biology you do not take **measurements** on organisms or other items, but classify them into different **categories**. For instance, birds belong to different species and have different colours; habitats (and Petri dishes) can have particular species present or absent; and people can be healthy or diseased. You cannot sensibly assign numbers to such arbitrarily defined classes; green is not larger in any real sense than yellow! For this reason you cannot use any of the statistical tests we examined in Chapters 4 and 5, which look for differences or associations between measurements.

Instead, this categorical data is best quantified by counting the numbers of observations in the different categories. This will allow you to estimate the **frequency** with which each character state turns up. This data can then be used to answer one of two questions:

- We might want to know whether the character frequencies in a single group are different from expected values. Do rats in a maze turn to the right rather than left at a different frequency from the expected 1 : 1? Or is the frequency of rickets different in a small mining town from that in the general population?
- We might want to know whether the character frequencies in two or more groups are different from each other. In other words, are certain characteristics associated with each other? For example, is smoking more common in men than women? Or do different insect species preferentially visit different species of flower?

6.2 The problem of variation

At first glance it might seem easy to tell whether character frequencies are different. When looking at a sample of sheep, if we found that eight were black and six white, we might conclude that black ones were commoner than white. Unfortunately, there might easily have been the same number of black and white sheep in the population and we might just have picked more black ones by chance.

A character state is, in fact, unlikely to appear at exactly the same frequency in a small sample as in the whole population. Let's examine what happens when we take samples of a population of animals, 50% of which are white and 50% black. In a sample of 2 there is only a 50% chance of getting a 1 : 1 ratio; the other times both animals would be either black or white. With 4 animals there will be a 1 : 1 ratio only 6 times out of 16; there will be a 3 : 1 or 1 : 3 ratio 4 times out of 16 and a 4 : 0 or 0 : 4 ratio once every 16 times.

As the number of animals in the sample increases, the most likely frequencies are those closer and closer to 1 : 1, but the frequency will hardly ever equal 1 : 1 exactly. In fact the probability distribution will follow an increasingly tight **binomial distribution** (Figure 6.1) with **mean** $\bar{x}$ equal to

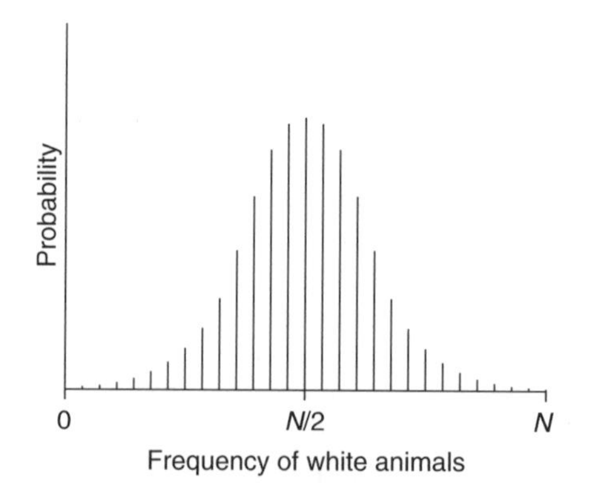

Figure 6.1 Binomial distribution. Probabilities of choosing different numbers of white animals in a sample of size *N* from a population with 50% white animals.

$N/2$, where N is the sample size and **standard deviation** s equal to $\sqrt{x}$. The probability that the ratio is near 1 : 1 increases, and the chances of it being further away decreases. However, there is always a finite, if increasingly tiny, chance of getting all white animals.

Things get more complex if the expected frequencies are different from 1 : 1 and if there are a larger number of categories, but essentially the same pattern will occur: as the sample size increases, the frequencies will tend to approach, but seldom equal, the frequencies in the population. The probability of obtaining frequencies similar to that of the population rises, but there is still a finite probability of the frequencies being very different.

So if the results from a sample are different from the expected frequency, you cannot be sure this is because the population you are sampling is really different. Even if you sampled 100 animals and all were white, the population still might have contained a ratio of 1 : 1; you might just have been very unlucky. However, the greater the difference and the larger the sample, the less likely this becomes. To determine whether differences from an expected frequency are likely to be real, you must use the final statistical test we will introduce, the **chi-squared test**. There are two main types of chi-squared (χ^2) test:

- The χ^2 test for differences
- The χ^2 test for association

6.3 The χ^2 test for differences

6.3.1 Purpose

To test whether character frequencies are different from expected values. It is best used when you expect numbers in different groups to be in particular ratios. Here are some examples:

- Experiments in Mendelian genetics
- Maze or choice chamber experiments
- Examining sex ratios
- Comparing your results with figures from the literature

6.3.2 Rationale

The test calculates the chi-squared statistic (χ^2); this is a measure of the difference between the observed frequencies and the expected frequencies. The larger χ^2, the less likely the results could have been obtained by chance if the population frequency was the expected one.

6.3.3 Carrying out the test

Steep 1: Formulate the null hypothesis
The null hypothesis is that the frequencies of the different categories in the population are equal to the expected frequencies.

Step 2: Calculate the test statistic

The χ^2 statistic is a measure of how far your observed frequencies are from the expected frequencies. It is given by the simple expression

$$\chi^2 = \sum \frac{(O - E)^2}{E} \tag{6.1}$$

where O is the observed frequency and E is the expected frequency for each character state. The larger the difference between the frequencies, the larger the value of χ^2 and the less likely it is that observed and expected frequencies are different just by chance. Similarly, the bigger the sample, the larger O and E, hence the larger the value of χ^2; this is because of the squared term in the top half of the fraction. So the bigger the sample you take, the more likely you will be to detect any differences.

The greater the number of possible categories, the greater the degrees of freedom; this also tends to increase χ^2. The distribution of χ^2 has been worked out for a range of degrees of freedom, and Table S3 (at the end of the book) gives the critical values of χ^2 above which there is less than a 5%, 1% or 0.1% probability of getting the observed values by chance. You cannot carry out this test in MINITAB.

Step 3: Calculate the significance probability

You must calculate the probability P of obtaining χ^2 values equal to or greater than the observed values if the null hypothesis were true. To do this you look up in Table S3 (at the end of the book) the critical value that χ^2 must exceed at $(N - 1)$ degrees of freedom, where N is the number of groups, for the probability to be less than 5%.

Step 4: Decide whether to reject the null hypothesis

- If χ^2 is greater than the critical value, you must reject the null hypothesis. You can say that the distribution is significantly different from expected.
- If χ^2 is less than the critical value, you cannot reject the null hypothesis. You have found no significant difference from the expected distribution.

Example 6.1

We will examine an example from Mendelian genetics in which F1 hybrids of smooth and wrinkled peas are crossed together. The following results were obtained:

Number of smooth peas = 69

Number of wrinkled peas = 31

We will test whether the ratio of smooth to wrinkled peas in the 100 progeny is different from the 3 : 1 ratio predicted by Mendelian genetics.

Formulate the null hypothesis

The null hypothesis is that the ratio smooth : wrinkled is 3 : 1.

Calculate the test statistic

The first thing to do is to calculate the expected values. Here 3/4 should be smooth and 1/4 wrinkled. Since there are 100 progeny,

$$\text{Expected number of smooth} = (3 \times 100)/4 = 75$$

$$\text{Expected number of wrinkled} = (1 \times 100)/4 = 25$$

So we have

$$\chi^2 = \sum \frac{(O - E)^2}{E}$$

$$\chi^2 = \frac{(69 - 75)^2}{75} + \frac{(31 - 25)^2}{25}$$

$$\chi^2 = 36/75 + 36/25 = 0.48 + 1.44 = 1.92$$

Calculate the significance probability

Looking up in Table S3 (at the end of the book) the value of χ^2 for $2 - 1 = 1$ degree of freedom, we find that χ^2 must be greater than 3.84.

Decide whether to reject the null hypothesis

We have $\chi^2 = 1.92 < 3.84$, so we have no evidence to reject the null hypothesis. The relative frequencies of smooth and wrinkled peas are not significantly different from the expected 3 : 1 ratio.

6.4 The χ^2 test for association

6.4.1 Purpose

To test whether the character frequencies of two or more groups are different from each other. In other words, to test whether character states are associated in some way. It is used when there is no expected frequency. Here are some examples:

- Ecological surveys: are different species found in different habitats?
- Medical surveys: are infection rates different for people in different blood groups?
- Sociological surveys: do men and women have a different probability of smoking?

6.4.2 Rationale

The test investigates whether the distribution is different from what it would be if the character states were distributed randomly among the population.

6.4.3 Carrying out the test

Just like the test for differences, this test follows all four of the usual steps of statistical tests.

We will examine an example from a sociological study which found that out of 30 men, 18 were smokers and 12 non-smokers, while of the 60 women surveyed, 12 were smokers and 48 were non-smokers. We will test whether the rates of smoking are significantly different between the sexes.

Step 1: Formulate the null hypothesis
The null hypothesis is that there is no difference between the frequencies of the groups, hence no association between the character states.

Step 2: Calculate the test statistic

Using a calculator
This is a complex process because before we can calculate χ^2 we must first calculate the expected values for each character state if there had been no association. The first stage is to arrange the data in a **contingency table**.

	Smoking	*Non-smoking*	*Total*
Men	18	12	30
Women	12	48	60
Total	30	60	90

It is now straightforward to calculate the frequencies if there had been no association between smoking and gender. Of the total number of people examined, one-third (30) were men, and one-third (30) of all people smoked. Therefore if the same proportion of men smoked as in the general population, you would expect one-third of all men (10) to be smokers. Hence 20 men should be non-smokers. Similarly, of the 60 women only one-third (20) should be smokers and 40 should be non-smokers.

A general expression for the expected number E in each cell of the contingency table is given by

$$E = \frac{\text{Column total} \times \text{Row total}}{\text{Grand total}} \tag{6.2}$$

where the grand total is the total number of observations (here 90). Therefore, the expected value for male smokers is found by multiplying its row total (30) by the column total (30) and dividing by 90, to give 10. These expected values are then put into the contingency table, written in parentheses. It is now straightforward to calculate χ^2 using equation (6.1).

	Smoking	*Non-smoking*	*Total*
Men	18 (10)	12 (20)	30
Women	12 (20)	48 (40)	60
Total	30	60	90

Using MINITAB
MINITAB can perform this whole process in an instant. Simply enter the observed frequencies into columns. From the Statistics menu choose Tables and then Chisquare Test. Enter the columns to be tested and run the test.

Step 3: Calculate the significance probability
To do this you look up the value that χ^2 must exceed at $(R-1) \times (C-1)$ degrees of freedom, where R is the number of rows in the table and C is the number of columns, for the probability to be less than 5%.

Step 4: Decide whether to reject the null hypothesis

- If χ^2 is greater than the critical value, you must reject the null hypothesis. You can say that the distribution is significantly different from expected, hence there is a significant association between the characters.
- If χ^2 is less than the critical value, you cannot reject the null hypothesis. You have found no significant difference from the expected distribution, hence no evidence of an association between the characters.

Example 6.2: Smoking data of Section 6.4.3

Formulate the null hypothesis
The null hypothesis is that men and women smoke with equal frequency. Therefore smoking is not associated with a particular sex.

Calculate the test statistic
MINITAB will produce the following output. Expected counts are printed below observed counts. Some versions of MINITAB also calculate P.

```
                    Men                 Women                 Total
   1               18                  12                      30
                   10.00               20.00
   2               12                  48                      60
                   20.00               40.00
Total              30                  60                      90
ChiSq =  6.400 + 3.200 + 3.200 + 1.600 = 14.400
df = 1
```

Calculate the significance probability
Looking up in Table S3 (at the end of the book) the value of χ^2 for $(2-1) \times (2-1) = 1$ degree of freedom, we find that χ^2 must be greater than 3.84.

Decide whether to reject the null hypothesis
We have $\chi^2 = 14.4 > 3.84$, so we have strong evidence to reject the null hypothesis. We can say there is a significant association between sex and smoking. In other words, the

two sexes are different in the frequency with which they smoke. In fact men smoke more than women.

You can tell even more about your results by looking at the χ^2 values for each of the cells. The larger the value, the more the results for the cell differ from the expected results. In this example χ^2 for male smokers is by far the largest at 6.4. Therefore we can say that in particular more men smoke than one would expect.

6.5 Validity of χ^2 tests

We have seen that the bigger the number of observations, the more likely you are to be able to detect differences or associations with the χ^2 test. In fact χ^2 tests are only valid if all expected values are larger than 5. If any expected values are lower than 5, there are two possibilities:

- You could combine data from two or more groups, but only if this makes biological sense. For instance, different species of fly could be combined in Problem 6.4 because flies have more in common with each other than with the other insects studied.
- If there is no sensible reason for combining data, small groups should be left out of the analysis.

6.6 Self-assessment problems

Problem 6.1

In an experiment to test the reactions of mice to a potential pheromone, they were run down a T-junction maze; the pheromone was released in one of the arms of the T. After the first 10 trials, 3 mice had turned towards the scent and 7 had turned away. After 100 trials, 34 had turned towards the scent and 66 had turned away. Is there any evidence of a reaction to the scent?

(a) After 10 trials
(b) After 100 trials

Problem 6.2

A cross was carried out between peas which were heterozygous in the two characters: height (tall H or short h) and pea colour (green G or yellow g). The following offspring were obtained.

	Number
Tall plants, green peas	87
Tall plants, yellow peas	34
Short plants, green peas	28
Short plants, yellow peas	11

For unlinked genes the expected ratios of each sort of plant are 9 : 3 : 3 : 1. Carry out a chi-squared test to determine whether there is any evidence of gene linkage between these characters.

Problem 6.3

A study of the incidence of a childhood illness in a small mining town showed that out of a population of 165 children, 9 had developed the disease. This compares with a rate of 3.5% in the country as a whole. Is there any evidence of a different rate in the town?

Problem 6.4

In a study of insect pollination, the numbers of insect visitors belonging to different taxonomic groups were investigated at flowers of different colours. The following results were obtained.

Insect visitors	*Flower colour*			*Total*
	White	*Yellow*	*Blue*	
Beetles	56	34	12	102
Flies	31	74	22	127
Bees and wasps	57	103	175	335
Total	144	211	209	564

(a) Carry out a χ^2 test to determine whether there is any association between the types of insects and the colour of the flowers they visit.
(b) Which cells have the three highest χ^2 values? What do these results tell you about the preferences of different insects?

Problem 6.5

A study was carried out to determine whether there is a link between the incidence of skin cancer and the possession of freckles. Of the 6045 people examined, 978 had freckles, of whom 33 had developed skin cancer. Of the remaining people without freckles, 95 had developed skin cancer. Is there any evidence that people with freckles have an increased risk of developing skin cancer?

Problem 6.6

A field study on the distribution of two species of newt found that of 745 ponds studied, 180 contained just smooth newts, 56 just palmate newts, 236 had both species present and the remainder had neither. Is there any association between the two species and, if so, what is it?

Choosing tests and designing experiments

How would you like your *t* test: Earl Grey or Darjeeling?

7.1 Introduction

The last few chapters have shown how to carry out statistical tests on the results of a range of experiments and surveys. But to become a successful research biologist you also need to be able to apply this knowledge correctly to design and analyse your own experiments, and to know when to ask for help. This chapter has three main aims:

- To point out the limitations of the information given in this small handbook.
- To enable you to work out which statistical test you should carry out on your results to answer particular questions.
- To enable you to design experiments or surveys which will generate the results needed to answer those questions.

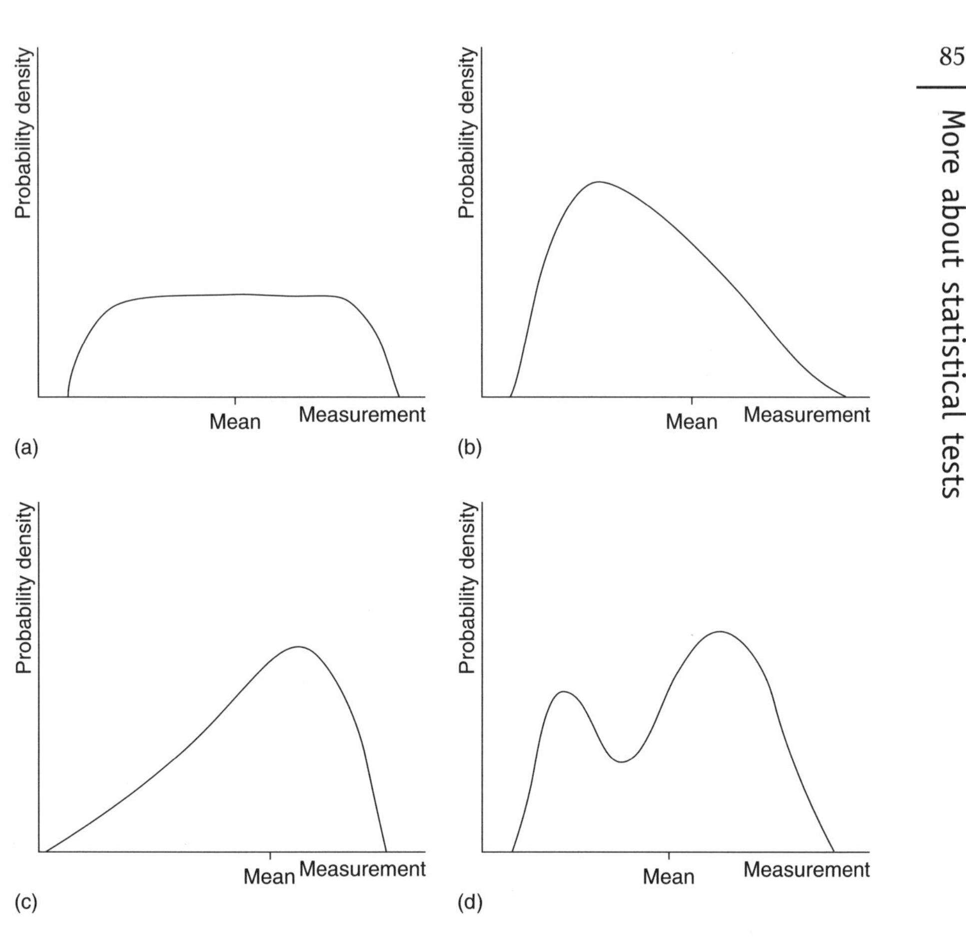

Figure 7.1 Non-normally distributed data. (a) A symmetrical distribution. (b) Positively skewed data. (c) Negatively skewed data. (d) Irregularly distributed data.

7.2 More about statistical tests

7.2.1 Dealing with non-normally distributed data

All the statistical tests which were introduced in Chapters 3 to 5 rely on the assumption that the data are normally distributed. Such tests as the t test, ANOVA, correlation and regression are known as **parametric tests**. However, data can also be distributed in other ways (Figure 7.1). In these cases parametric tests are not always valid. The data items may have to be transformed to make them more normally distributed, or you may even have to carry out a different **non-parametric test** which makes no assumptions about the distribution. Once the data have been transformed you can then simply carry out the appropriate parametric test.

Symmetrically distributed data

If the data are not normally distributed but are symmetrically distributed around the mean (Figure 7.1a) and the sample size is over 20, it is generally valid to use parametric tests. This is because the means of samples tend to be normally distributed if the sample size is large enough.

Skewed data

If the data are asymmetrically distributed, either being **positively skewed** (Figure 7.1b) or **negatively skewed** (Figure 7.1c), you may need to use a **transformation** before you do the statistical analysis.

Positively skewed data are commonly found when examining natural populations because there tend to be many more young (and hence small) organisms in a population. Positively skewed data may be transformed into a distribution which is closer to the normal distribution in several ways, all of which separate smaller numbers. Possible transformations of a variable x are to take $\sqrt{x}$, $\log_{10} x$, or $1/x$.

Negatively skewed data are less common. To transform negatively skewed data into normally distributed data, you need to separate the larger numbers. Possible transformations are powers of x like x^2, x^3 or even higher powers of x.

Proportional data

Proportional data, such as the ratio of root to total plant mass, cannot be normally distributed because values cannot fall below 0 or rise above 1. Each data point x is transformed by calculating $\arcsin\sqrt{x}$.

Small irregularly distributed samples and ranked data

If you need to deal with small samples of data distributed in an irregular way (Figure 7.1d), as is commonly the case with many ecological and behavioural surveys, there is no sensible transformation you can use to make the data normally distributed. This is also true of the ranked data (like pecking order or degree of patterning) which may be all that ecologists, behavioural scientists or psychologists can collect.

In these cases you cannot use parametric tests. Instead, you need to use an appropriate **non-parametric test**. Table 7.1 lists the most commonly used non-parametric tests along with the parametric test for which they are the alternative. For details of how to carry out these tests, you should refer to larger statistical textbooks like Sokal and Rohlf's *Biometry*. However, if you have the chance you should always use parametric tests because they are more powerful; there is more chance of detecting significant effects.

7.2.2 Investigating the shape of distributions

To investigate whether you can assume data are normally distributed and, if not, to work out exactly what the distribution is, you could carry out a range of tests.

Table 7.1 Parametric tests and their non-parametric equivalents.[a]

Parametric test	*Non-parametric equivalent*
One-sample t test	One-sample sign test
Paired t test	Wilcoxon matched pairs test
Two-sample t test	Mann–Whitney U test
One-way ANOVA	Kruskal–Wallis test
Correlation	Rank correlation
Regression	Rank correlation

[a] Non-parametric tests are used either when samples are small and data are irregularly distributed or when examining rank data.

Descriptive statistics

The simplest thing to do is to carry out descriptive statistics. If your data are normally distributed, about two-thirds of points should be within one standard deviation from the mean, and all points within three standard deviations.

Box and whisker plots

A box and whisker plot is a very useful way of determining whether or not your distribution is symmetrical. To do this you need to rank the observations in ascending order and calculate five values: the lowest and highest values; the **median** or value of the middle point (or average of the two middle points if the sample size is an even number); and the **lower and upper quartiles** (Q1 and Q3), the values exceeded by 25% and 75% of the data points respectively. Statistical packages like MINITAB will calculate these values for you when they calculate the descriptive statistics, as we saw in Chapter 3.

These points (or MINITAB) can then be used to draw the box and whisker plot (Figure 7.2a). To draw a box and whisker plot in MINITAB, go to the Statistics menu, choose EDA and then Boxplot. Finally, enter the column you want to examine.

The box and whisker plot in Figure 7.2a is for the Problem 3.4 data about the mass of newborn babies. This plot is approximately symmetrical about the median, showing that the distribution is approximately symmetrical. If you have an asymmetric plot, you should transform it using the appropriate transformation from Section 7.2.1. You can then examine the box and whisker plot of the transformed data to determine whether it is symmetrical. If so, your transformation has been successful. If not, you must try a different transformation.

Stem and leaf plots

To quickly investigate the shape of the distribution, MINITAB will also allow you to produce a stem and leaf plot (Figure 7.2b). Simply go to the Statistics menu, choose EDA and then Stem-and-Leaf. Finally, enter the column you want to examine.

LLYFRGELL COLEG MENAI LIBRARY
BANGOR GWYNEDD LL57 2TP

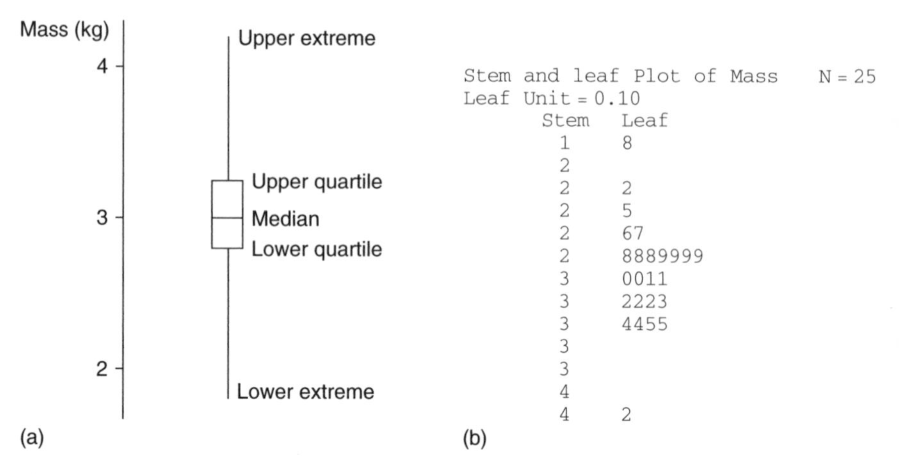

Figure 7.2 Pictorial presentation. (a) A box and whisker plot shows that the data are distributed fairly symmetrically. (b) A stem and leaf plot shows that the data are distributed approximately normally. The data are the masses of newborn babies from Problem 3.4.

The plot is like a sideways-on histogram. The stem shows the common component of each number (here units of 1 kg); the leaf shows the individual component (here units of 0.1 kg).

Tests for normality

There are also several more formal statistical tests for normality, such as the Kolmogorov–Smirnoff test (see Sokal and Rohlf's *Biometry*), but there is no room to describe them here.

7.2.3 Complex statistical tests

In addition to the non-parametric tests, there are a whole range of more complex tests which have been developed by statisticians to answer a wide range of questions.

Investigating differences caused by many factors

We have seen how one-way and two-way ANOVA can be used to detect the effect of one or two factors. However, using multiway ANOVA, it is also possible to investigate several factors simultaneously.

Investigating associations between several variables

We have seen how correlation and regression can be used to investigate linear relationships between two sets of data. Other techniques can be used to investigate more complex relationships and those between three or more variables. These include multiple regression and principal component analysis (PCA).

Investigating both differences and associations
A useful technique which can be used to compare the slopes and intercepts of regression lines is analysis of covariance (ANCOVA). But perhaps the ultimate statistical technique is the general linear model (GLM), a development of ANOVA, regression and ANCOVA, which can be used to simultaneously investigate differences and associations.

7.3 Choosing tests

Once you are clear in your own mind what has been measured and what you are trying to find out, it is very straightforward to choose which statistical test to use. All you have to do is follow the sequence of questions in Figure 7.3 (also shown on the inside cover) until you reach the correct test.

There is only one complication. On the left-hand side of the chart, the final box may have two alternative tests: a parametric test in bold type and an equivalent non-parametric test in medium type. Unless it is not valid, you are recommended to use the parametric test; this will be more powerful. Here are some examples.

Example 7.1

You are comparing the seed weights of four varieties of winter wheat, and you have weighed 50 seeds of each variety. Which test should you use?

Solution

- You have taken a *measurement* of weight, so you should choose the left-hand option.
- You are looking for *differences* between the varieties, so you should choose the left-hand option.
- You have *more than two* samples, so you should choose the right-hand option.
- You are investigating the effect of only *one factor*, variety, so you should choose the left-hand option.
- You are examining a *continuous* measurement, weight, likely to be normally distributed and with a sample size of well over 20, so you should choose the boldface option.
- You should use the parametric test *one-way ANOVA*.

Example 7.2

You are investigating how the swimming speed of fish depends on their length, and you have measured both for 30 fish. Which test should you use?

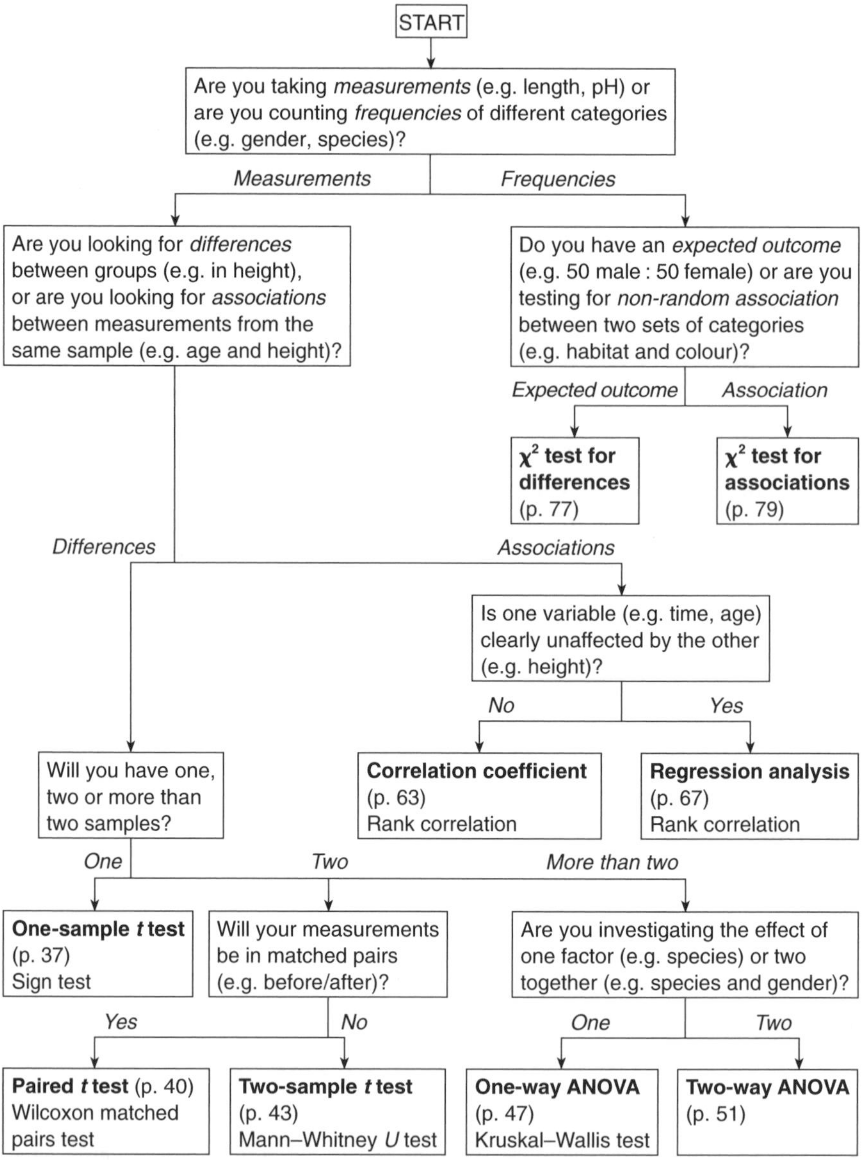

START

Are you taking *measurements* (e.g. length, pH) or are you counting *frequencies* of different categories (e.g. gender, species)?

- *Measurements* → Are you looking for *differences* between groups (e.g. in height), or are you looking for *associations* between measurements from the same sample (e.g. age and height)?
  - *Differences* → Will you have one, two or more than two samples?
    - *One* → **One-sample *t* test** (p. 37) Sign test
    - *Two* → Will your measurements be in matched pairs (e.g. before/after)?
      - *Yes* → **Paired *t* test** (p. 40) Wilcoxon matched pairs test
      - *No* → **Two-sample *t* test** (p. 43) Mann–Whitney *U* test
    - *More than two* → Are you investigating the effect of one factor (e.g. species) or two together (e.g. species and gender)?
      - *One* → **One-way ANOVA** (p. 47) Kruskal–Wallis test
      - *Two* → **Two-way ANOVA** (p. 51)
  - *Associations* → Is one variable (e.g. time, age) clearly unaffected by the other (e.g. height)?
    - *No* → **Correlation coefficient** (p. 63) Rank correlation
    - *Yes* → **Regression analysis** (p. 67) Rank correlation
- *Frequencies* → Do you have an *expected outcome* (e.g. 50 male : 50 female) or are you testing for *non-random association* between two sets of categories (e.g. habitat and colour)?
  - *Expected outcome* → **χ^2 test for differences** (p. 77)
  - *Association* → **χ^2 test for associations** (p. 79)

Figure 7.3 Decision chart for statistical tests. Start at the top and follow the questions down until you reach the appropriate box. The tests in medium type are non-parametric equivalents for irregularly distributed or ranked data.

Solution

- You have taken *measurements* on speed and length, so you should choose the left-hand option.
- You are looking for an *association* between the two measurements, speed and length, so you should choose the right-hand option.
- One measurement, length, is clearly *unaffected by* the other, speed, so you should choose the right-hand option.
- You should carry out *regression analysis*.

Example 7.3

You are investigating the incidence of measles in children resident in hospitals and comparing it with the national average; you have surveyed 540 children. Which test should you use?

Solution

- You have counted the *frequencies* of children in two *categories* (with and without measles), so you should choose the right-hand option.
- You are comparing your sample with an *expected outcome*, so you should choose the left-hand option.
- You should calculate χ^2 *for differences*.

Example 7.4

You are investigating the relationship between the weight and social rank of domestic hens, and you have observed 34 birds. Which test should you use?

Solution

- You have taken *measurements* on weight and rank, so you should choose the left-hand option.
- You are looking for an *association* between the two measurements, weight and social rank, so you should choose the right-hand option.
- The measurements are *not unaffected by* each other – weight can affect rank, and rank can also affect weight – so you should choose the left-hand option.
- Since social rank is by definition *rank* data you should choose the option in medium type.
- You should carry out *rank correlation*.

7.4 Designing experiments

The role of experiments and surveys in biology is to help you answer questions about the natural world. According to scientific papers, this is amazingly simple. All you need do is to examine and take measurements on small numbers of organisms or cells. The results you obtain can then be subjected to statistical analysis to determine whether differences between groups or associations between variables are real or could have occurred by chance. Nothing could appear simpler, but much time can be wasted carrying out badly designed experiments or surveys. Many experiments which are carried out could never work; in others the size of the sample is either too small to detect the sorts of effects which might be expected or much larger than necessary. Still other experiments are ruined by a confounding variable.

Before you carry out an experiment or survey, you should therefore ask yourself the following questions:

- Could it possibly work?
- How large should my samples be?
- How can I exclude confounding variables?

Most of the work which a scientist does is aimed at answering these questions and is carried out *before* the final experiment is performed. Scientific papers therefore contain a wholly distorted account of the process of science and conceal the hard work involved in the planning process. The key to success in experimentation is careful forward planning and preliminary work. The main techniques are preparation, replication, randomisation and blocking.

7.4.1 Preparation

To find out whether an experiment could work, you must first find out something about the system you are studying by reading the scientific literature, by carrying out the sort of rough calculations we examined in Chapter 2, or by making preliminary examinations yourself.

Example 7.5

Consider an experiment to test whether the lead from petrol fumes affects the growth of roadside plants. There would be no point in carrying out an experiment if we already knew from the literature that lead only reduces growth at concentrations above 250 ppm, while levels measured at our roadside were only 20 ppm.

7.4.2 Replication

The key to working out how many **replicates** you need in your samples is knowledge of the system you are examining. You must have enough replicates to allow statistical analysis to tease apart the possible effects from random variability, but not much more else you would be wasting your time. Therefore you need to know two things: the size of the effects you want to be able to detect and the variability of your system. In general, the smaller the effect you want to detect and the greater the variability, the larger the sample sizes you need. For various tests it is relatively straightforward to work out how many replicates you need to test for differences.

One-sample and paired *t* tests

As we saw in Chapter 4, in one-sample and paired *t* tests the smallest detectable difference *D* occurs when *t* is around about 2. In other words, when *D* is around twice the standard error $\overline{\mathrm{SE}}$:

$$D \approx 2\overline{\mathrm{SE}}$$

But standard error is the standard deviation divided by the square root of the number in the sample ($\overline{\mathrm{SE}} = s/\sqrt{N}$). Therefore

$$D \approx 2s/\sqrt{N}$$

The number of replicates you need *N* can therefore be found by rearranging the equation, so that

$$N \approx 4(s/D)^2 \qquad (7.1)$$

Before carrying out your experiment you should therefore strive to obtain figures for the size of the effect you want to detect and the likely standard deviation of samples. You could do this by looking up values from the literature or by carrying out a small pilot experiment.

Example 7.6

A survey is to be carried out to determine whether tortoises on a volcanic island are larger than their mainland relatives. It is known that the masses of mainland animals have a standard deviation *s* equal to 30% of the mean weight. Assuming island tortoises show the same variation, how many would be needed to detect (a) a 20% and (b) a 10% difference in mass?

Solution

(a) $n \approx 4 \times (30/20)^2 \approx 9$

(b) $n \approx 4 \times (30/10)^2 \approx 36$

Two-sample *t* tests and ANOVA

The smallest detectable difference *D* in two-sample *t* tests also occurs when *t* is around 2. In this case, however, *D* is twice the standard error of the difference $\overline{\mathrm{SE}}_d$, which itself is around 1.5 times the standard error of each sample. So in this case

$$D \approx 2\overline{\mathrm{SE}}_d \approx 2(1.5\overline{\mathrm{SE}}) \approx 3\overline{\mathrm{SE}}$$

and using $\overline{\mathrm{SE}} = s/\sqrt{N}$ as we did above, this gives the expression

$$N \approx 9(s/D)^2 \qquad (7.2)$$

Therefore to detect a given smallest difference D with a two-sample t test, you need a larger value of N than with a one-sample t test. The same applies to ANOVA.

Example 7.7

Tortoises from two different volcanic islands are to be compared. Assuming the standard deviation of each population's mass is 30 g, how many tortoises should be sampled from each island to detect (a) a 10 g difference and (b) a 10 g difference in mean mass?

Solution

(a) $N \approx 9 \times (30/20)^2 \approx 20$

(b) $N \approx 9 \times (30/10)^2 \approx 81$

Chi-squared tests for differences

If you have an expected outcome for the frequency of a particular category, it is once again straightforward to work out the sample size you require. As we saw in Chapter 6, the standard error $\overline{\mathrm{SE}}$ for the number of times a category will come up is the square root of the expected value E. The smallest difference D that will be statistically significant is around twice the standard error, so

$$D \approx 2\sqrt{E}$$

But D equals the proportional difference d multiplied by N. Similarly, E is the expected proportion e multiplied by N. Therefore the equation can be rewritten as

$$Nd = 2\sqrt{Ne}$$

Rearranging this equation we find that

$$N \approx 4e/d^2 \qquad (7.3)$$

Example 7.8

We want to know how many people in a town we need to test to detect whether there is a 1% difference in the incidence of hair lice from the national figure of 4%.

Solution

Inserting values for e of 0.04 and for d of 0.01 into the equation, we get $N \approx 4 \times 0.04/0.01^2 \approx 1600$.

Tests for associations

It is much more difficult to determine before you have done the experiment the sample size you need to test for **associations**. All you can do is to examine your data as you collect them and test them as you go along. In general, the more data you have the better.

7.4.3 Randomisation and blocking

The final key to designing successful experiments is to exclude confounding variables; you should alter only the factor whose effect you want to examine, and keep all other factors the same.

Clearly this is straightforward for some factors. You can easily arrange an experiment so that the replicates are all kept at the same temperature, at the same light levels and over the same period of time. However, it is not possible to keep every factor the same; for instance, you cannot grow plants in exactly the same position as each other, or in exactly the same soil, and you cannot harvest and test them at exactly the same time. In such cases, where differences are inevitable, systematic errors between samples can be avoided by using **randomisation** techniques. You should randomise the position of plants and the order in which they are tested using random numbers from tables or computers, so that none of the groups is treated consistently differently from the others.

Randomisation seems straightforward, but it can lead to a rather uneven spread of replicates. For instance, if a field plot is very long and thin or split up between several fields, you might easily get different numbers of replicates of each sample at different ends or in different fields (Figure 7.4a). In this case you should split the plot into several blocks (Figure 7.4b) and randomise the same number of replicates from each sample within each block. The same should be done if testing or harvesting of your experiment takes several days. You should test an equal number of replicates from each sample each day. This use of **blocking** has the added advantage that, if you analyse your data using ANOVA, you can also investigate separately the differences between the blocks.

Field 1 (a)

A	B	B	A
A	A	A	B

Field 2 (a)

B	B	A	A
B	B	A	B

(a)

Field 1 (b)

A	B	B	A
B	A	A	B

Field 2 (b)

A	A	B	A
B	B	A	B

(b)

Figure 7.4 Blocking can help to avoid confounding variables: an agricultural experiment with two treatments, each with eight replicates. (a) The treatments have been randomised, but this has produced an uneven distribution of treatments, both between the fields (more of treatment A is in field 1) and within them (in field 2 treatment B is concentrated at the left-hand side). (b) The treatments have been randomised within blocks (two applications of each treatment are made to each side of each field); this removes the possible confounding variable of position.

7.5 Dealing with results

During your experiment you should look at your results as soon as possible, preferably while you are collecting them, and try to think what they are telling you about the natural world. Calculate the mean and standard deviations of measurements, plot the results on a graph, or look at the frequencies in different categories. Once you can see what seems to be happening, you should write down your ideas in your laboratory book, think about them and then tell your supervisor or a colleague. Do not put your results into a spreadsheet and forget about them until the write-up.

Only after you have worked out what you think is happening should you carry out your statistical analysis to see if the trends you identified are significant. Usually, if a trend is not obvious to the naked eye it is unlikely to be significant. So always use statistics as a tool; do not allow it to become the master!

7.6 Presenting results

Once you have worked out which effects are significant, you should present your results in a way that emphasises the biological reality of what you have shown. Do not just comment on whether effects were significant, but mention how large they were. Consider the statement

Dogs were significantly larger than cats ($t = 6.43$, $P < 0.001$).

This does not tell you how much larger. It is better to say something like

The height of plants which were given extra nitrogen was around 40% greater than those which had not ($t = 3.45$, $P < 0.01$).

You could also give 95% confidence limits. Similarly, you should give the equations of regression lines, e.g.

Age a had a significant positive effect on density d; the equation of the regression line is $d = 124 + 0.32a$ ($P < 0.05$).

Again, confidence intervals would be useful.

7.7 Concluding remarks

No book of 100 pages can tell you all you need to know about using mathematics and statistics. This book merely scratches the surface of the subject. However, if you can master the contents of this book, you will probably have gained enough knowledge about data handling and statistics for your undergraduate and maybe even for your postgraduate studies. Certainly, as a biologist who has carried out biomechanical and ecological research for some 15 years, I have rarely if ever used any statistical test not included here. The key to dealing with data in biology is to keep within your limitations. You don't need to know a huge amount, but you must be able to apply what you do know sensibly. Indeed, simplicity in thinking is often an advantage. If you restrict yourself to straightforward mathematics and statistics, other scientists are more likely to follow what you have done!

7.8 Self-assessment problems

Problem 7.1

A clinician wants to find out if there is any link between energy intake (in calories) and heart rate in old people. She collects data on both of them from 150 volunteers. Which statistical test should she choose to determine whether energy intake and heart rate are linked?

Problem 7.2

An ecologist collects data about the numbers of individuals that belong to five species of crow feeding in three different habitats: farmland, woodland and mountain. How will he analyse his data to determine whether different crows are distributed non-randomly in different habitats?

Problem 7.3

A doctor wants to find out if there is any difference in insulin levels between three races of people: Afro-Caribbean, Asian and Caucasian. Having collected data on insulin levels from 30 people of each race, which statistical test should he use to answer his question?

Problem 7.4

A genetics student wants to find out whether two genes are linked: one for shell colour (brown dominant, yellow recessive) and one for having a banded or plain shell (banded dominant, plain recessive). To do this, she crosses two pure-bred lines of brown banded snails with yellow plain snails. The result is an F1 generation, all of which are brown and banded. These are crossed to produce an F2 generation. What statistical test should she perform to test whether there is in fact any linkage?

Problem 7.5

A new medication to lower blood pressure is being tested in field trials. Forty patients were tested before and after taking the drug. Which test should the clinicians use to best determine whether it is having an effect?

Problem 7.6

It has been suggested that pot plants can help the survival of patients in intensive care by providing the room with increased levels of oxygen from photosynthesis. Carry out a rough calculation to work out if this idea is worth testing. (*Hint*: estimate how fast the plant is growing and hence laying down carbohydrates and exporting oxygen.)

Problem 7.7

In an investigation into starch metabolism in mutant potatoes, the effect of deleting a gene is investigated. It is expected that this will reduce the level to which starch builds up. Large numbers of previous experiments have shown that the mean level of starch in ordinary potatoes is 21 *M* with standard deviation 7.9 *M*. Assuming the standard deviation in mutants is similar to that in ordinary potatoes, how many replicates would have to be examined to detect a significant difference in mutants whose mean starch level is 16 *M*, a 5 *M* fall?

Problem 7.8

An experiment is being designed to test the effect of shaking maize plants on their extension growth. There will be two groups: shaken plants and unshaken controls. It is known that maize usually has a mean height of 1.78 m with standard deviation of 0.36 m. How many replicates of experimental and control plants must be grown to detect a height difference of 0.25 m?

Problem 7.9

The national rate of breast cancer is given as 3.5% of women over 45 years old. It has been suggested that silicone implants may increase the rate of such cancers. How many women with implants would need to be tested to detect a doubling of the risk?

Problem 7.10

Design an experiment to test the relative effects of applying four different amounts of nitrogen fertiliser, 0, 3.5, 7 and 14 g of nitrogen per square metre per year, in 25 fortnightly applications (researchers get a fortnight off at Christmas) on to chalk grassland at Wardlaw, Derbyshire. The field site is split into 16 plots, each of dimensions 1 m × 1 m in an 8 m × 2 m grid (see diagram). You are supplied with a quantity of 20×10^{-3} M ammonium nitrate solution fertiliser. Mark the different plots and describe exactly what you would apply to each plot.

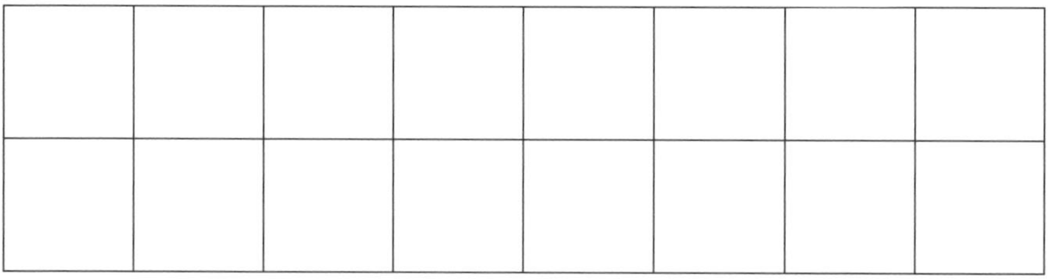

Glossary

ANOVA Abbreviation for analysis of variance: a widely used test which can determine whether there are significant differences between groups or associations between them.

association A numerical link between two sets of measurements.

binomial distribution The pattern by which the sample frequencies in two groups tends to vary.

blocking A method of eliminating confounding variables by spreading replicates of the different samples evenly among different blocks.

category A character state which cannot meaningfully be represented by a number.

causal relationship Relationship between two variables whereby one affects the other but is not itself affected.

chi-squared (χ^2) A statistical test which determines whether there are differences between real and expected frequencies in one set of categories, or associations between two sets of categories.

confidence limits Limits between which estimated parameters have a defined likelihood of occurring. It is common to calculate 95% confidence limits, but 99% and 99.9% confidence limits are also used. The range of values between the upper and lower limits is called the confidence interval.

confounding variables Variables which invalidate an experiment if they are not taken into account.

contingency table A table showing the frequencies of two sets of character states, which allows you to calculate expected values in a chi-squared test for association.

correlation A statistical test which determines whether there is linear association between two sets of measurements.

critical values Tabulated values of test statistics; if the absolute value of a calculated test statistic is greater than the appropriate critical value, the null hypothesis must be rejected.

data Observations or measurements you have taken (your results) which are used to work things out about the world.

degrees of freedom (DF) A concept used in parametric statistics, based on the amount of information you have when you examine samples. The number of degrees of freedom is generally the total number of observations you make minus the number of parameters you estimate from the samples.

dependent variable A variable in a regression which is affected by another variable.

descriptive statistics Statistics which summarise the distribution of a single set of measurements.

distribution The pattern by which a measurement or frequency varies.

error bars Bars drawn upwards and downwards from the mean values on graphs; error bars can represent the standard deviation or the standard error.

estimate A parameter of a population which has been calculated from the results of a sample.

exponent A power of 10 which allows large or small numbers to be readily expressed and manipulated.

exponential relationship A relationship which follows the general equation $y = ae^{bx}$. If $b > 0$ this is exponential growth; if $b < 0$ this is exponential decay.

frequency The number of times a particular character state turns up.

Imperial Obsolete system of units from the United Kingdom.

independent variable A variable in a regression which affects another variable but is not itself affected.

interaction A synergistic or inhibitory effect between two factors which can be picked up by using two-way ANOVA.

intercept The point where a straight line crosses the y-axis.

logarithm to base 10 ($\log_{10}$) A function of a variable y such that if $y = 10^x$ then $x = \log_{10} y$.

mean (μ) The average of a population. The estimate of μ is called $\bar{x}$.

mean square The variance due to a particular factor in analysis of variance (ANOVA).

measurement A character state which can meaningfully be represented by a number.

median The central value of a distribution (or average of the middle points if the sample size is even).

metric Units based on the metre, second and kilogram but not necessarily SI.

natural logarithm ($\log_e$ or ln) A function of a variable y such that if $y = e^x$ then $x = \log_e y$ or $\ln y$.

normal distribution The usual pattern for measurements which are influenced by large numbers of factors.

null hypothesis A preliminary assumption in a statistical test that the data shows no differences or associations. The test then works out the chances of the null hypothesis being correct.

parametric test A statistical test which assumes data are normally distributed.

population A potentially infinite group on which measurements could be taken. Parameters of populations usually have to be estimated from the results of samples.

post hoc tests Statistical tests carried out if an analysis of variance is significant; they are used to determine which groups are different.

power relationship A relationship which follows the general equation $y = ax^b$.

prefix A multiple or divisor of 1000 which allows large or small numbers to be readily expressed.

quartiles Upper and lower quartiles are values exceeded by 25% and 75% of the data points, respectively.

rank Numerical order of a data point.

regression A statistical test which analyses how one set of measurements is affected by another.

replicates The individual data points.

replication The use of large numbers of measurements to allow one to estimate population parameters.

sample A subset of a possible population on which measurements are taken. These can be used to estimate parameters of the population.

scatter plot A point graph between two variables which allows one to visually determine whether they are associated.

scientific notation A method of representing large or small numbers, giving them as a number between 1 and 10 multiplied by a power of 10.

SI Système International: the common standard of units used in modern science based on the metre, second and kilogram.

significance probability The chances that a certain set of results could be obtained if the null hypothesis were true.

significant difference A difference which has less than a 5% probability of having happened by chance.

skewed data Data with an asymmetric distribution.

slope The gradient of a straight line.

standard deviation (σ) A measure of spread of a group of measurements: the amount by which on average they differ from the mean. The estimate of σ is called s.

standard error (SE) A measure of the spread of sample means: the amount by which they differ from the true mean. Standard error equals standard deviation divided by the square root of the number in the sample. The estimate of SE is called $\overline{\mathrm{SE}}$.

standard error of the difference ($\overline{\mathrm{SE}}_d$) A measure of the spread of the difference between two estimated means.

***t* distribution** The pattern by which sample means of a normally distributed population tend to vary.

***t* tests** Statistical tests which analyse whether there are differences between measurements on a single population and an expected value, between measurements on the same population, or between measurements on two populations.

transformation A mathematical function used to make the distribution of data more symmetrical and so make parametric tests valid.

two-tailed tests Tests which ask merely whether observed values are different from an expected value or each other, not whether they are larger or smaller.

variance A measure of the variability of data: the square of their standard deviation.

Further reading

Mathematics and statistics are huge areas of knowledge, so this short book has of necessity been superficial and selective in the material it covers. For more background and for more information about particular aspects of the use of numbers in biology, the reader is referred to the following titles. The books vary greatly in their approach and in the degree of mathematical competence required to read them, so you should choose your texts carefully. Sokal and Rohlf (1994) is really the bible of biological statistics, but most students will struggle with its high mathematical tone. If you really struggle with the basics of maths, you are best referred to Croft and Davison (1995) or Burton (1998). A very simple introduction to statistical thinking (without any equations!) is Rowntree (1981). Excellent advice about even complex experimental design can be found in Heath (1995) and Wardlaw (1985), who also reproduce several more useful statistical tables. General advice and examples in problem solving can be found in Ennos and Bailey (1995), while Jones, Reed and Weyers (1998) provide useful advice over a wide range of biology.

BURTON, R. F. (1998) *Biology by Numbers*. Cambridge University Press, Cambridge.

CROFT, A. and DAVISON, R. (1995) *Foundation Mathematics*. Longman, Harlow.

EASON, G., COLES, C. W. and GETTINBY, G. (1992) *Mathematics and Statistics for the Bio-Sciences*. Ellis Horwood, Chichester.

ENNOS, A. R. and BAILEY, S. E. R. (1995) *Problem Solving in Environmental Biology*. Longman, Harlow.

HEATH, D. (1995) *An Introduction to Experimental Design and Statistics for Biology*. UCL Press, London.

JONES, A., REED, R. and WEYERS, J. (1998) *Practical Skills in Biology*, 2nd edn. Longman, Harlow.

ROWNTREE, D. (1981) *Statistics without Tears*. Penguin Books, London.

SOKAL, R. R. and ROHLF, F. J. (1994) *Biometry*, 3rd edn. W. H. Freeman, San Francisco.

WARDLAW, A. C. (1985) *Practical Statistics for Experimental Biologists*. John Wiley, New York.

WATT, T. A. (1993) *Introductory Statistics for Biology Students*. Chapman & Hall, London.

Solutions

Chapter 2

Problem 2.1

(a) m^2
(b) $m\ s^{-1}$ (though the number will obviously be very low!)
(c) m^{-3} (number per unit volume)
(d) no units (it's one concentration divided by another)

Problem 2.2

(a) 192 MN or 0.192 GN
(b) 102 μg or 0.102 mg
(c) 0.12 ms (120 μs would imply that you had measured to 3 significant figures)
(d) 213 mm or 0.213 m

Problem 2.3

(a) 4.61×10^{-5} J
(b) 4.61×10^{8} s

Problem 2.4

(a) 3.81×10^{9} Pa
(b) 4.53×10^{-3} W
(c) 3.64×10^{-1} J
(d) 4.8×10^{-6} kg (remember that the SI unit of mass is the kg)
(e) 2.1×10^{-16} kg (remember that the SI unit of mass is the kg)

Problem 2.5

(a) $250 \times 10^3 = 2.50 \times 10^5$ kg
(b) $0.3 \times 10^5 = 3 \times 10^4$ Pa
(c) $24 \times 10^{-10} = 2.4 \times 10^{-9}$ m

Problem 2.6

In each case use a similar degree of precision as the original measurements.

(a) $35 \times 0.9144 = 32.004$ m
$= 32$ m (2 significant figures)
(b) $(3 \times 0.3048) + (3 \times 2.54 \times 10^{-2}) = 0.9144 + 0.0762$
$= 0.99$ m (2 significant figures)
(c) $9.5 \times (0.9144)^2 = 7.943\ \text{m}^2$
$= 7.9\ \text{m}^2$ (2 significant figures)

Problem 2.7

(a) $(1.23 \times 2.456) \times 10^{(3+5)}\ \text{m}^2 = 3.02 \times 10^8\ \text{m}^2$ (3 significant figures)
(b) $(2.1/4.5) \times 10^{(-2+4)}\ \text{J kg}^{-1} = 0.4666 \times 10^2$
$= 4.7 \times 10^1\ \text{J kg}^{-1}$ (2 significant figures)

Problem 2.8

(a) 1.3 mmol
(b) 365 MJ or 0.365 GJ
(c) 0.24 μm (not 240 nm, because this implies knowledge to 3 significant figures)

Problem 2.9

The concentration is the number of cells divided by the volume in which they were found. The dimensions of the box are 1×10^{-3} m by 1×10^{-3} m by 1×10^{-4} m. Therefore its volume is $1 \times 10^{(-3-3-4)} = 1 \times 10^{-10}\ \text{m}^3$. The concentration of blood cells is therefore

$$652/(1 \times 10^{-10}) = (6.52 \times 10^2)/(1 \times 10^{-10})$$
$$= (6.52/1) \times 10^{(2+10)}$$
$$= 6.52 \times 10^{12}\ \text{m}^{-3}$$

Problem 2.10

The volume of water which had fallen is the depth of water which had fallen multiplied by the area over which it had fallen.

$$\text{Depth} = 0.6 \times 2.54 \times 10^{-2}$$
$$= 1.524 \times 10^{-2}\ \text{m}$$

$$\text{Area} = 2.6 \times 10^4 \text{ m}^2$$

$$\text{Therefore} \quad \text{Volume} = 1.524 \times 10^{-2} \times 2.6 \times 10^4$$
$$= 3.962 \times 10^2 \text{ m}^3$$
$$= 4 \times 10^2 \text{ m}^3 \quad \text{(1 significant figure)}$$

Problem 2.11

The concentration is the number of moles of glucose per litre.

$$\text{Number of moles} = \frac{\text{Mass in grams}}{\text{Molecular mass}}$$

$$= \frac{25}{(6 \times 12) + (12 \times 1) + (6 \times 16)}$$

$$= 25/180 = 1.3888 \times 10^{-1}$$

$$\text{Concentration} = 1.3888 \times 10^{-1}/2 = 6.9444 \times 10^{-2}\ M$$
$$= 6.9 \times 10^{-2}\ M \quad \text{(2 significant figures)}$$

Problem 2.12

The first thing to work out is the volume of CO_2 that was produced.

$$\text{Volume } CO_2 \text{ produced} = \text{Volume of air} \times \text{Proportion of it which is } CO_2$$
$$= 45 \times 0.036 = 1.62 \text{ litres}$$

And we know that at room temperature and pressure 1 mol of gas takes up 24 litres, so

$$\text{Number of moles } CO_2 = 1.62/24 = 6.75 \times 10^{-2}$$

The mass of CO_2 produced equals the number of moles multiplied by the mass of each mole of CO_2. Since the mass of 1 mol of $CO_2 = 12 + (2 \times 16) = 44$ g, we have

$$\text{Mass of gas produced} = 6.75 \times 10^{-2} \times 44 = 2.97 \text{ g}$$
$$= 2.97 \times 10^{-3} \text{ kg}$$

Production of this gas took 5 minutes $= 5 \times 60 = 300$ s, so

$$\text{Rate of gas production} = (2.97 \times 10^{-3})/300$$
$$= 9.9 \times 10^{-6} \text{ kg s}^{-1}$$

Problem 2.13

The energy produced by the reaction was converted to heat, and heat energy = mass × specific heat × temperature rise. First, we need to work out the mass of water. Fortunately, this is easy as 1 litre of water weighs 1 kg. Therefore 0.53 litre has a mass of 0.53 kg. From Table 2.5 we can see that water has a specific heat of 4.2×10^3 J K^{-1} kg^{-1}, therefore

$$\begin{aligned} \text{Heat energy} &= 0.53 \times 4.2 \times 10^3 \times 2.4 \text{ J} \\ &= 5342.4 \text{ J} \\ &= 5.3 \times 10^3 \text{ J or } 5.3 \text{ kJ} \quad \text{(2 significant figures)} \end{aligned}$$

Problem 2.14

The first thing to do is work out the number of moles of X you will use:

$$\begin{aligned} \text{Number of moles} &= \text{Volume (in litres)} \times \text{Concentration (in moles per litre)} \\ &= (8 \times 80 \times 10^{-3}) \times 3 \times 10^{-3} \\ &= 1.92 \times 10^{-3} \end{aligned}$$

Now obtain the mass of 1.92×10^{-3} mol of X:

$$\begin{aligned} \text{Mass (in grams)} &= \text{Number of moles} \times \text{Molecular mass} \\ &= 1.92 \times 10^{-3} \times 258 \\ &= 0.495 \text{ g} \end{aligned}$$

And finally the cost of 0.495 g of X:

$$\begin{aligned} \text{Cost} &= \text{Number of grams} \times \text{Price per gram} \\ &= 0.495 \times 56 \\ &= £28 \quad \text{(2 significant figures)} \end{aligned}$$

Since this is far less than £1000 you will easily be able to afford it.

Problem 2.15

The first thing to do is to work out the volume of methane produced by bogs per year. We have

$$\text{Yearly production} = \text{Daily productivity} \times \text{Area of bog} \times \text{Days in a year}$$

And we have

$$\text{Daily productivity} = 21 \text{ ml m}^{-2} = 2.1 \times 10^{-2} \text{ l m}^{-2}$$

$$\text{Area of bogs} = 3.4 \times 10^6 \text{ km}^2 = 3.4 \times 10^{12} \text{ m}^2$$

Therefore

$$\begin{aligned} \text{Yearly production} &= 2.1 \times 10^{-2} \times 365 \times 3.4 \times 10^{12} \\ &= 2.606 \times 10^{13} \text{ litres} \end{aligned}$$

Next you need to work out how many moles this is equal to and hence the mass of methane produced per year. Since 1 mol of gas takes up 24 litres, we have

$$\text{Number of moles} = \frac{\text{Volume in litres}}{24}$$

$$= (2.606 \times 10^{13})/24$$

$$= 1.086 \times 10^{12}$$

We also have

Mass of methane (in grams) = Number of moles × Molecular mass

Since the molecular mass of methane is $12 + (1 \times 4) = 16$, therefore

$$\text{Mass of methane} = 1.086 \times 10^{12} \times 16$$

$$= 1.737 \times 10^{13}\ \text{g}$$

$$= 1.737 \times 10^{10}\ \text{kg}$$

However, since this is three times as effective as CO_2, this is equivalent to $1.737 \times 10^{10} \times 3 = 5.2 \times 10^{10}$ kg of CO_2.

How does this compare with the amount of CO_2 produced by burning fossil fuels? This equals 25 Gt. We need to convert to kg:

$$25\ \text{Gt} = 25 \times 10^{9}\ \text{t} = 25 \times 10^{12}\ \text{kg} = 2.5 \times 10^{13}\ \text{kg}$$

This is much more. The ratio of the effect of fossil fuel to the effect of bog methane production is

$$\frac{2.5 \times 10^{13}}{5.2 \times 10^{10}} = \sim 500$$

Therefore bog methane will have a negligible effect compared with our use of fossil fuels.

Problem 2.16

(a) 1.65
(b) 2.65
(c) –3.35
(d) 6
(e) 0

Problem 2.17

(a) 25.1
(b) 251
(c) 3.98×10^{-4}
(d) 10^4
(e) 1

Problem 2.18

(a) In 3×10^{-4} M HCl the concentration of H^+ is $[H^+] = 3 \times 10^{-4}$ M. Therefore pH = 3.5
(b) In 4×10^{-6} M H_2SO_4 the concentration of H^+ is $[H^+] = 8 \times 10^{-6}$ M. Therefore pH = 5.1

Problem 2.19

Concentration of H^+ ions in pH $2.1 = 10^{-2.1} = 7.94 \times 10^{-3}$ M. But each molecule of H_2SO_4 has two hydrogen ions. Therefore the concentration of $H_2SO_4 = (7.94 \times 10^{-3})/2 = 3.97 \times 10^{-3}$ M.

$$\text{Number of moles} = \text{Concentration} \times \text{Volume}$$
$$= 3.97 \times 10^{-3} \times 0.160$$
$$= 6.35 \times 10^{-4}$$

$$\text{Molecular mass of } H_2SO_4 = 2 + 32 + 64 = 98$$

$$\text{Mass of } H_2SO_4 = \text{Number of moles} \times \text{Molecular mass (in grams)}$$
$$= 6.35 \times 10^{-4} \times 98$$
$$= 6.22 \times 10^{-2} \text{ g}$$
$$= 6.2 \times 10^{-5} \text{ kg} \quad \text{(2 significant figures)}$$

Problem 2.20

(a) 3.40
(b) −3.73
(c) 0

Problem 2.21

(a) 20.1
(b) 0.050
(c) 1

Chapter 3

Problem 3.1

95% will have heart rates between $75 \pm (1.96 \times 11)$, i.e. between 53 and 97 beats per minute.

Problem 3.2

Mean = 5.71 g, $s = 0.33$ g

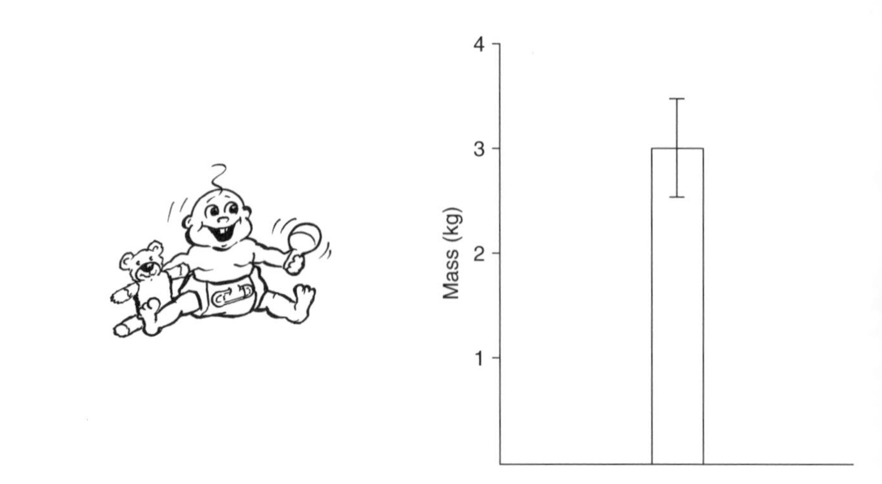

Figure A1 Mean birthweight. Error bar represents standard deviation.

Problem 3.3

(a) Mean = 5.89, $s = 0.31$, $\overline{\mathrm{SE}} = 0.103$, 95% CI = 5.89 ± (2.306 × 0.103) = 5.65 to 6.13.
(b) Mean = 5.95, $s = 0.45$, $\overline{\mathrm{SE}} = 0.225$, 95% CI = 5.95 ± (3.182 × 0.225) = 5.23 to 6.67. The 95% confidence interval is three times wider than for (a).

Problem 3.4

(a) Mean (s) = 3.00 (0.47) kg, $n = 25$. $\overline{\mathrm{SE}} = 0.093$ kg
(b) The bar chart is shown in Figure A1.

Chapter 4

Problem 4.1

The null hypothesis is that the mean score = 58%. The mean score of the students is $\bar{x} = 58.36$ with $s = 13.70$ and $\overline{\mathrm{SE}} = 2.74$. The score seems higher than expected but in the one-sample t test, $t = (58.36 - 58)/2.74 = 0.13$ to 2 decimal places. The absolute value of t, 0.13, is therefore well below the value of 2.064 needed for significance at 24 degrees of freedom. The probability for this happening by chance is therefore well over 0.05 (in fact it is 0.90). Therefore students did not perform significantly differently from expected.

Problem 4.2

The null hypothesis is that the mean mass of tomatoes is 50 g. Looking at descriptive statistics, $\bar{x} = 44.1$ g, $s = 8.6$ g, $\overline{\mathrm{SE}} = 2.15$ kg. This seems well below 50 g. In the one-sample t test to determine whether the mean mass is significantly different from 50 g, $t = (44.1 - 50)/2.15 = -2.74$ to 2 decimal places. The absolute value of t, 2.74, is well above the value of 2.131 required for significance at 15 degrees of freedom (remember that is the magnitude of t and not whether it is positive or negative that matters). The probability of this happening by chance is therefore well under 0.05 (in fact it is 0.016). Therefore the tomatoes are significantly lighter than the expected 50 g. The 95% confidence interval for mass is $44.1 \pm (2.131 \times 2.15) = 39.5$ to 48.7 g.

Problem 4.3

The null hypothesis is that students' scores after the course were the same as before it. The mean score was 58.1 before and 53.8 after. The scores seem to be worse afterwards, but to find whether the difference is significant you need to carry out a paired t test. This shows a mean difference $d = -4.3$, $s = 5.7$ and $\overline{SE}_d = 1.79$. Therefore $t = -4.3/1.79 = -2.40$ to 2 decimal places. Its absolute value, 2.40, is larger than the value of 2.306 needed for significance at $9 - 1 = 8$ degrees of freedom. P is therefore less than 0.05 (0.04 in fact). Therefore the course did have a significant effect. After the course most students got worse marks!

The 95% confidence interval for the difference is $-4.3 \pm (2.306 \times 1.79) = -0.02$ to -8.4.

Problem 4.4

(a) The null hypothesis is that pH was the same at dawn and dusk. The two-sample t test carried out on MINITAB gives the following results:

```
TWOSAMPLE T FOR dawn VS dusk
          N        MEAN        STDEV        SE MEAN
dawn      12       5.542       0.705        0.20
dusk      11       6.445       0.641        0.19

95 PCT CI FOR MU dawn - MU dusk: (-1.49, -0.32)
Pooled Standard Deviation = 0.675
TTEST MU dawn = MU dusk (VS NE): T = -3.21 P = 0.0042 DF = 21
```

We can see that the mean pH at dawn (5.542) is well below the mean at dusk (6.445) but is the difference significant? MINITAB calculates $t = -3.21$ and $P = 0.0042$, well below 0.05. Therefore it is clear we must reject the null hypothesis and conclude there is a significant difference between pH at dawn and dusk; in fact it's higher at dusk. MINITAB also calculates that the 95% confidence interval for the difference is -1.49 to -0.32.

(b) You cannot use a paired t test, because the cells which you measured are not identifiably the same. Indeed, different numbers of cells were examined at dawn and dusk.

Problem 4.5

The null hypothesis is that the control and supported plants have the same mean yield. The two-sample t test carried out on MINITAB gives the following results:

```
TWOSAMPLE T FOR Control VS Support
                 N          MEAN          STDEV        SE MEAN
Control         20         10.28           1.60          0.36
Support         20         10.06           1.55          0.35

95 PCT CI FOR MU Control - MU Support: (-0.79, 1.22)
Pooled Standard Deviation = 1.58
TTEST MU Control = MU Support (VS NE): T = 0.43 P = 0.67 DF = 38
```

We can see that the supported plants have a slightly lower mean yield (10.06 g) than control plants (10.28 g), but is this difference significant? MINITAB calculates $t = 0.43$ and $P = 0.67$, well above 0.05. Therefore we have no evidence to reject the null hypothesis and we have found no significant difference in yield between control and supported plants. The 95% confidence interval for mean yield are –0.79 to 1.22 g.

Problem 4.6

The null hypothesis is that the mean diameters of the colonies in the four samples are the same. MINITAB gives the following results:

```
ANALYSIS OF VARIANCE
SOURCE    DF      SS       MS       F      P
FACTOR     3    0.055   0.018   0.08   0.973
ERROR     50   12.126   0.243
TOTAL     53   12.181
                                   INDIVIDUAL 95 PCT CI'S FOR MEAN
                                   BASED ON POOLED STDEV
LEVEL      N    MEAN     STDEV    -+---------+---------+---------+-----
Control   14   5.0643   0.5048     (------------*------------)
A         14   5.1500   0.4768         (-------------*--------------)
B         10   5.1300   0.4945      (---------------*--------------)
C         16   5.1125   0.4938       (------------*-----------)
                                  -+---------+---------+---------+-----
POOLED STDEV = 0.4925             4.80      5.00      5.20      5.40
```

Here the means are hardly different, the error bars all overlap and F is very small (0.08). The significance probability is very high ($P = 0.973 > 0.05$). Therefore it is clear there is no significant difference between the treatments, so the antibiotics do not appear to have any significant effect.

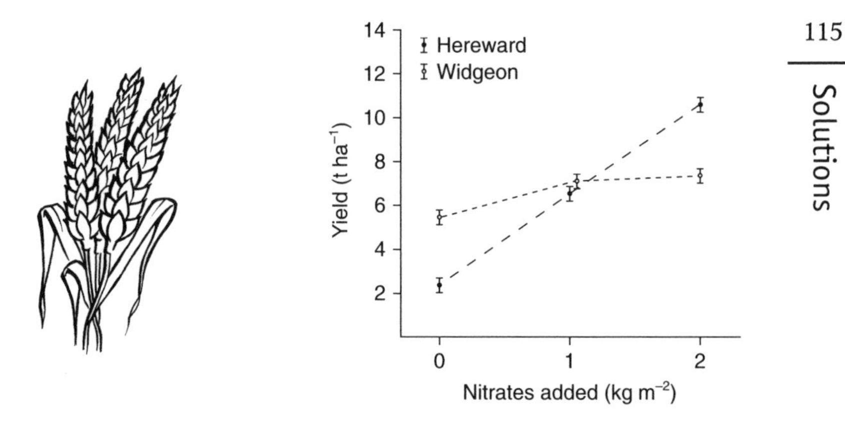

Figure A2 Mean yields. Mean yields ($\overline{\text{SE}}$) of two varieties of wheat at three different nitrogen concentrations (for each group $n = 9$). Increasing nitrogen clearly raises yield more for Hereward than for Widgeon.

Problem 4.7

Looking at the degrees of freedom (DF), it is clear that $4 + 1 = 5$ groups must have been examined, and in total $29 + 1 = 30$ observations must have been made. F is quite small (171) and P is high ($0.35 > 0.05$). Therefore there was no significant difference between the groups.

Problem 4.8

(a) Looking at the main table, it is clear the nitrogen level and the interaction term are both significant ($P < 0.05$). In contrast, variety has no effect ($P > 0.05$).

(b) Looking at the results and plotting the descriptive statistics on a graph, it is clear that adding nitrogen increases yield in both varieties (Figure A2). This is the cause of the significant term for nitrogen level. The average yield of the two varieties is about the same (the cause of the non-significant variety term), but nitrogen has more effect on yield in Hereward than in Widgeon (the cause of the significant interaction term). Hence Widgeon does better without nitrogen; Hereward does better with lots of nitrogen.

Chapter 5

Problem 5.1

(a) Plot cell number against time.

(b) Plot chick pecking order against parent pecking order, since the behaviour and health of chicks is more likely to be affected by their parents than vice versa.

(c) Plot weight against height, because weight is more likely to be affected by height than vice versa.

(d) You can plot this graph either way, because length and breadth are a measure of size and may both be affected by the same factors.

Problem 5.2

(a) If $\log_{10} A = 0.3 + 2.36 \log_{10} L$, taking inverse logarithms gives

$$A = 10^{0.3} \times L^{2.36}$$

$$A = 2.0L^{2.36}$$

(b) If $\log_e N = 2.3 + 0.1T$, taking inverse natural logarithms gives

$$N = e^{2.3} \times e^{0.1T}$$

$$N = 10e^{0.1T}$$

Problem 5.3

The null hypothesis is that there is no linear association between leaf area and stomatal density. In the correlation analysis, MINITAB calculates that the correlation coefficient $r = -0.944$. This looks like a strong negative correlation, but is it significant? To be significant at $10 - 2 = 8$ degrees of freedom, the absolute value of r must be greater than 0.632. This is true, so the effect is significant. Leaf area and stomatal density show a significant negative correlation.

Problem 5.4

(a) Bone density is the dependent variable, so it should be plotted along the vertical axis (Figure A3).

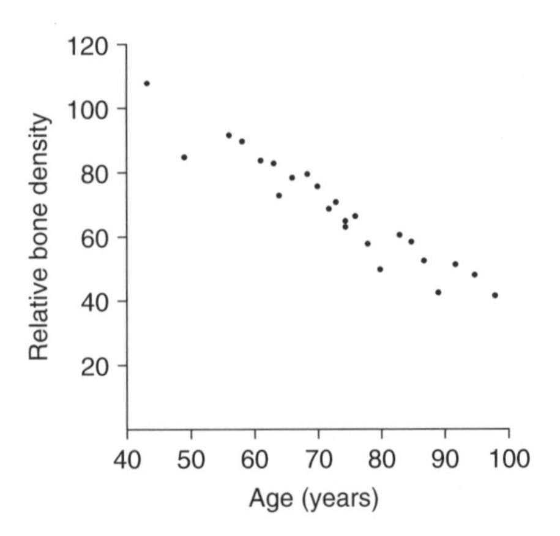

Figure A3 Post-menopausal women: relative bone density versus age.

(b) In MINITAB the regression analysis yields the following results:


```
THE REGRESSION EQUATION IS
Density = 151 - 1.13 Age
Predictor        Coef          Stdev        t ratio       P
Constant      151.277          5.637          26.83     0.000
Age            -1.12751        0.07577       -14.88     0.000
s = 5.167          R-sq = 91.0%          R-sq(adj) = 90.6%

ANALYSIS OF VARIANCE
SOURCE        DF       SS         MS          F          P
Regression     1     5911.3     5911.3     221.41     0.000
Error         22      587.4       26.7
Total         23     6498.6
```

From the graph and the equation of the regression line, it appears that bone density falls significantly with age. To determine whether the fall is significant, we must examine the first set of output. The value of the t ratio for the effect of age = −14.88 and the probability of this happening by chance is 0.000, which is less than 0.05. Therefore the slope is significantly different from 0. We can say there is a significant change of bone density with age (in fact a loss).

(c) Expected density at age 70 is found by inserting the value of 70 into the regression equation:

$$\text{Density} = 151.277 - (1.127\,51 \times 70) = 72.3$$

Problem 5.5

(a) Worm zinc concentration depends on environmental zinc concentration, not vice versa, so worm zinc concentration must be plotted along the y-axis (Figure A4).

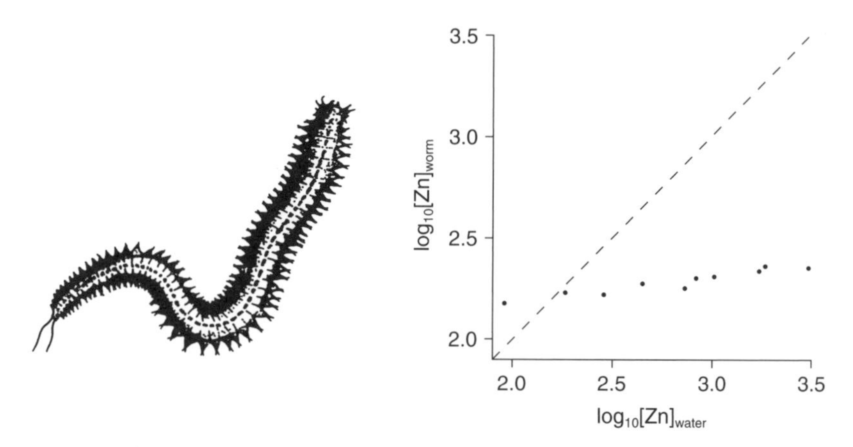

Figure A4 Relationship between $\log_{10}[Zn]_{water}$ and $\log_{10}[Zn]_{worm}$. The dashed line shows the expected relationship if worms did not regulate [Zn].

(b) In MINITAB the regression analysis comes up with the following results:

```
THE REGRESSION EQUATION IS
log10 [Zn] Worm = 1.94 + 0.119 log10 [Zn] Water
Predictor            Coef       Stdev      t ratio      P
Constant             1.94451    0.03342    58.19        0.000
log10 [Zn] Water     0.11920    0.01168    10.21        0.000
s = 0.01723              R-sq = 92.9%      R-sq(adj) = 92.0%

ANALYSIS OF VARIANCE
SOURCE        DF     SS          MS          F         P
Regression    1      0.030916    0.030916    104.17    0.000
Error         8      0.002374    0.000297
Total         9      0.033290
```

It is clear that the zinc concentration in the worms does increase with the zinc concentration in the water, but the slope is much lower than 1, being only 0.119 20. To investigate whether the slope is significantly different from 1 we must test the null hypothesis that the actual slope equals 1. To do this we carry out the following t test:

$$t = \frac{\text{Actual slope} - \text{Expected slope}}{\text{Standard deviation of slope}}$$

Here $t = (0.119\,20 - 1)/0.011\,68 = -75.41$. Its absolute value, 75.41, is much greater than the value of 2.306 needed for a significant effect at $10 - 2 = 8$ degrees of freedom. Therefore the slope is significantly different from 1 (less in fact). It is clear that the worms must be actively controlling their internal zinc concentrations.

Problem 5.6

(a) The relationship between seeding rate and yield is shown in Figure A5. It looks as if the yield rises to a maximum at a seeding rate of 300 m^{-2}, before falling again.

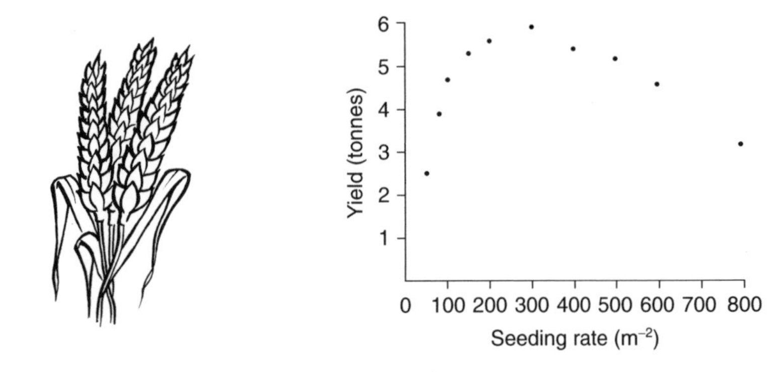

Figure A5 Relationship between seeding rate and yield.

(b) In MINITAB the regression analysis yields the following results:

```
THE REGRESSION EQUATION IS
Yield = 4.66 - 0.00009 Seeding Rate

10 CASES USED 3 CASES CONTAIN MISSING VALUES
Predictor          Coef          Stdev        t ratio        P
Constant         4.6587         0.6186          7.53       0.000
Seeding rate   -0.000090        0.001555       -0.06       0.955
s = 1.175              R-sq = 0.0%            R-sq(adj) = 0.0%

ANALYSIS OF VARIANCE
SOURCE          DF        SS          MS          F          P
Regression       1       0.005       0.005       0.00       0.955
Error            8      11.036       1.380
Total            9      11.041
```

It is clear that the regression equation explains essentially none of the variability ($r^2 = 0\%$), and the slope is not significantly different from 0 ($P = 0.955$, which is much greater than 0.05).

(c) There is no significant linear relationship between seeding rate and yield; the relationship is curvilinear. The moral of this exercise is that linear relationships are not the only ones you can get, so it is important to examine the data graphically.

Problem 5.7

(a) The logged data are shown in Figure A6. $\log_{10}$(jaw width) is plotted along the vertical axis because it is the factor which is more likely to depend on the other.

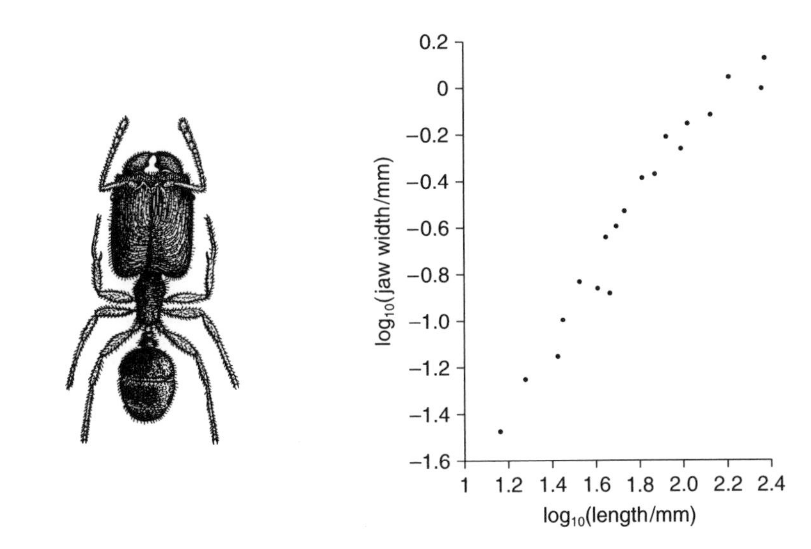

Figure A6 Relationship between $\log_{10}$(length) and $\log_{10}$(jaw width) of army ants.

(b) In MINITAB the regression analysis gives the following results:

```
THE REGRESSION EQUATION IS
log10(jaw width) = -1.36 + 1.49 log10(length)
Predictor              Coef        Stdev      t ratio        P
Constant           -1.36194      0.05180      -26.29     0.000
log10(jaw width)    1.48542      0.06610       22.47     0.000
s = 0.04022           R-sq = 96.6%          R-sq(adj) = 96.4%

ANALYSIS OF VARIANCE
SOURCE          DF        SS          MS          F          P
Regression       1     0.81670     0.81670     504.94     0.000
Error           18     0.02911     0.00162
Total           19     0.84581

UNUSUAL OBSERVATIONS
Obs.    C3        C4         Fit Stdev.Fit      Residual  St.Resid
8     0.724  -0.36653  -0.28609  0.00952     -0.08044    -2.06R
R denotes an obs. with a large st. resid.
```

Converting back to real numbers:

$$w = 10^{-1.36} l^{1.49} = 0.0436 l^{1.49}$$

where w is jaw width and l is body length.

(c) The slope of the regression line, 1.485 42, looks larger than 1, but is this effect significant? We need to carry out a t test, as in Problem 5.5, with the null hypothesis that the slope = 1:

$$t = \frac{\text{Actual slope} - \text{Expected slope}}{\text{Standard deviation of slope}}$$

Here $t = (1.485\,42 - 1)/0.0661 = 7.34$ to 2 decimal places. This is much larger than the value of 2.101 needed for significance at $20 - 2 = 18$ degrees of freedom. Therefore it is clear that the slope is significantly different from 1. The jaws actually get relatively wider in bigger ants.

Chapter 6

Problem 6.1

The null hypothesis is that the mice are equally likely to turn to the right as to the left. Therefore the expected ratio is 1 : 1.

(a) After 10 trials, expected values are 5 towards the scent and 5 away. $\chi^2 = (-2)^2/5 + 2^2/5 = 0.80 + 0.80 = 1.60$ to 2 decimal places. This is below the critical value of 3.84 needed for significance for 1 degree of freedom, so there is as yet no evidence of a reaction.

(b) After 100 trials, expected values are 50 towards the scent and 50 away. $\chi^2 = (-16)^2/50 + 16^2/50 = 5.12 + 5.12 = 10.24$. This is greater than the critical value of 3.84 needed for significance for 1 degree of freedom, so there is clear evidence of a reaction. The mice seem to avoid the scent, in fact.

This problem shows the importance of taking large samples, as this will improve the chances of detecting effects.

Problem 6.2

The null hypothesis is that there is no linkage, so in this sample of 160 plants the expected numbers in each class are 90, 30, 30 and 10. Therefore $\chi^2 = (-3)^2/90 + 4^2/30 + (-2)^2/30 + 1^2/10 = 0.100 + 0.533 + 0.133 + 0.100 = 0.87$ to 2 decimal places. This is below the critical value of 7.34 needed for significance for 3 degrees of freedom, so there is no evidence of a ratio different from 9 : 3 : 3 : 1 and so no evidence of linkage.

Problem 6.3

The null hypothesis is that the incidence of illness in the town was the same as for the whole country. The expected values for illness in the town = 3.5% of 165 = 5.8, and therefore 159.2 without the illness. $\chi^2 = (9 - 5.8)^2/5.8 + (156 - 159.2)^2/159.2 = 3.2^2/5.8 + (-3.2)^2/159.2 = 1.76 + 0.06 = 1.82$. This is below the critical value of 3.84 needed for significance for 1 degree of freedom, so there is no evidence of a different rate of illness.

Problem 6.4

The null hypothesis is that the insects are randomly distributed about different coloured flowers. The completed table with expected values is given here, and it shows that in many cases the numbers of insects on flowers of particular colours is very different from expected. But is this a significant association? A χ^2 test is needed.

Insect visitors	*Flower colour*			*Total*
	White	*Yellow*	*Blue*	
Beetles	56	34	12	102
	(26.04)	(38.16)	(37.80)	
Flies	31	74	22	127
	(32.43)	(47.51)	(47.06)	
Bees and wasps	57	103	175	335
	(85.53)	(125.33)	(124.14)	
Total	144	211	209	564

(a) Working with 4 decimal places for the expected values, e.g. 26.0426 instead of 26.04, we obtain

$$\chi^2 = 34.461 + 0.453 + 17.608 + 0.063 +$$
$$= 14.767 + 13.346 + 9.518 + 3.978 + 20.837$$
$$= 115.031 = 115.03 \quad \text{(2 decimal places)}$$

This value is higher than the critical value of 9.48 needed for $(3-1) \times (3-1) = 4$ degrees of freedom. Therefore we can conclude there is a significant association between insect type and flower colour.

(b) The highest χ^2 values are 34.46 for beetles and white flowers; 20.84 for bees and wasps and blue flowers; and 17.61 for beetles and blue flowers. Looking at the values, more beetles are found at white flowers than expected, so beetles in particular favour them; similarly, bees and wasps favour blue flowers; but beetles seem to avoid blue flowers.

Problem 6.5

The null hypothesis is that people with and without freckles have the same incidence of cancer. The completed contingency table is as shown and seems to indicate that the number of people with freckles who have cancer is greater than expected.

	Healthy	*Cancer*	*Total*
Freckles	945 (957.3)	33 (20.7)	978
No freckles	4981 (4968.7)	95 (107.3)	5076
Total	5926	128	6054

But is this a significant effect? A χ^2 test gives the following result: working with 4 decimal places for the expected values, we obtain

$$\chi^2 = 0.159 + 7.343 + 0.031 + 1.415 = 8.948$$
$$= 8.95 \quad \text{(2 decimal places)}$$

This value is higher than the critical value of 3.84 needed for 1 degree of freedom. Therefore there is a significant association between possession of freckles and skin cancer. More people with freckles get the disease.

Problem 6.6

The first thing to do is to calculate the number of ponds with neither newt species. This equals $745 - (180 + 56 + 236) = 273$. The null hypothesis is that there is no association between the presence in ponds of smooth and palmate newts. The completed contingency table is shown here and appears to indicate that there are far more ponds than expected with both species or with neither species present.

	− Smooth	*+ Smooth*	*Total*
− Palmate	273 (200.05)	180 (252.95)	453
+ Palmate	56 (128.95)	236 (163.05)	292
Total	329	416	745

Working with expected values to 4 decimal places, a χ^2 test gives the following results:

$$\chi^2 = 26.602 + 21.039 + 41.270 + 32.639 = 121.550$$
$$= 121.55 \quad \text{(2 decimal places)}$$

The value is higher than the critical value of 3.84 needed for $(2-1) \times (2-1) = 1$ degree of freedom. Therefore there is a significant association between the presence of the two species. In fact, the newts seem to be positively associated with each other. When one species is present, it is more likely that the other species will be present as well.

Chapter 7

Problem 7.1

Correlation. She is looking at measurements, looking for an association between two sets of measurements, and neither variable is clearly independent of the other.

Problem 7.2

χ^2 test for association. He is looking at frequencies in different categories, and looking for an association between two types of category (species and habitat).

Problem 7.3

One-way ANOVA. He is looking at measurements, looking for differences between groups; there are more than two groups, but only one type of factor (race).

Problem 7.4

χ^2 test for differences. She is dealing with frequencies in different categories, and there are expected frequencies (9 : 3 : 3 : 1).

Problem 7.5

The paired t test. They are looking at measurements, looking for differences, they will compare only two groups (before and after), and they have identifiable people to pair.

Problem 7.6

The net production by plants of oxygen via photosynthesis results in them growing. Therefore if we can estimate how much a pot plant grows, we can estimate how much oxygen it produces. Let's suppose it grows at the (fast) rate of 1 g dry mass per day (so after a year it would have a wet weight of over a kilogram).

Now oxygen is produced by the reaction

$$6CO_2 + 6H_2O \rightarrow C_6H_{12}O_6 + 6O_2$$

But 1 mol of glucose weighs $(12 \times 6) + 12 + (16 \times 6) = 180$ g, so the number of moles of glucose produced per day $= 1/180 = 5.556 \times 10^{-3}$. For every 1 mol of dry matter produced, 6 mol of O_2 is also produced. Therefore for every 1 g of dry mass produced by the plant, the number of moles of oxygen produced $= 6/180 = 3.333 \times 10^{-2}$.

Since 1 mol of oxygen takes up 24 litres, this makes up a volume of $3.333 \times 10^{-2} \times 24 = 0.80$ litres $= 0.8 \times 10^{-3}$ m^3 $= 8 \times 10^{-4}$ m^3. How does this compare with the amount of oxygen in the room. Well, let's imagine a room of $5\text{ m} \times 4\text{ m} \times 2.5\text{ m}$ high, containing 20% oxygen. The volume of oxygen $= 5 \times 4 \times 2.5 \times 0.2 = 10$ m^3. This is over 10 000 times greater. The tiny contribution of the plant will be far too small to make a difference. There is no point in doing the experiment.

Problem 7.7

The number of replicates required $\approx 4 \times (7.9/5)^2 \approx 10$. To play safe, it would be best to use something over this number, say around 15–20.

Problem 7.8

Number $\approx 9 \times (0.36/0.25)^2 \approx 19$. To play safe, it would be sensible to use around 25 replicates for the experimental plants and another 25 replicates for the controls.

Problem 7.9

A doubling of the risk means an increase of 0.035. Therefore $N \approx (4 \times 0.035)/0.035^2 \approx 115$. To be on the safe side, one would probably test over 150 people.

Problem 7.10

The first thing to do is to arrange for replication in your experiment. Each treatment should be given to four plots. Next you must decide how to arrange the treatments around the plots. You could randomise totally, arranging treatments randomly in each of the 16 plots. However, in this case one treatment might tend to be restricted to one end of the site. A better solution is to split the site into four 2 m × 2 m blocks and randomise each of the four treatments within each block (see diagram).

0	3.5	7	14	3.5	14	0	7
7	14	3.5	0	0	7	3.5	14

Next you must calculate how much fertiliser to apply to each plot. A litre of 1 *M* ammonium nitrate will contain 1 mol of the substance. The formula of ammonium nitrate is NH_4NO_3, so this will contain 2 mol of nitrogen, a mass of 28 g (the relative atomic mass of nitrogen is 14). Therefore the mass of nitrogen in 1 litre of 20×10^{-3} *M* ammonium nitrate fertiliser is given by

$$\text{Mass N} = 1 \times 0.020 \times 28 = 0.56 \text{ g}$$

To supply 14 g the volume required = 14/0.56 = 25 litres.

Each plot has area 1 m^2 so must be supplied with 25 × 1 = 25 litres per year. But this must be spread over 25 applications, so the volume which must be applied each visit is 25/25 = 1 litre of fertiliser.

What about the other plots? You could apply 0.5, 0.25 and 0 litres of fertiliser to get the correct rates of 7, 3.5 and 0 g of nitrogen per year. However, you would be adding different quantities of water to each plot! To control for this, you should add 1 litre of fertiliser diluted by a factor of 2 and 4 respectively to the 7 and 3.5 g plots, and 1 litre of water to each zero nitrogen plot.

Statistical tables

Table S1: Critical values for the t statistic

Critical values of t at the 5%, 1% and 0.1% significance levels. Reject the null hypothesis if the absolute value of t is larger than the tabulated value at the chosen significance level, for the calculated number of degrees of freedom.

Degrees of freedom	*Significance level*		
	5%	*1%*	*0.1%*
1	12.706	63.657	636.619
2	4.303	9.925	31.598
3	3.182	5.841	12.941
4	2.776	4.604	8.610
5	2.571	4.032	6.859
6	2.447	3.707	5.959
7	2.365	3.499	5.405
8	2.306	3.355	5.041
9	2.262	3.250	4.781
10	2.228	3.169	4.587
11	2.201	3.106	4.437
12	2.179	3.055	4.318
13	2.160	3.012	4.221
14	2.145	2.977	4.140
15	2.131	2.947	4.073
16	2.120	2.921	4.015
17	2.110	2.898	3.965
18	2.101	2.878	3.922
19	2.093	2.861	3.883
20	2.086	2.845	3.850
21	2.080	2.831	3.819
22	2.074	2.819	3.792
23	2.069	2.807	3.767
24	2.064	2.797	3.745
25	2.060	2.787	3.725
26	2.056	2.779	3.707
27	2.052	2.771	3.690
28	2.048	2.763	3.674
29	2.045	2.756	3.659
30	2.042	2.750	3.646
40	2.021	2.704	3.551
60	2.000	2.660	3.460
120	1.980	2.617	3.373
∞	1.960	2.576	3.291

Table S2: Critical values for the correlation coefficient *r*

Critical values of the correlation coefficient *r* at the 5%, 1% and 0.1% significance levels. Reject the null hypothesis if your absolute value of *r* is greater than the tabulated value at the chosen significance level, for the calculated number of degrees of freedom.

Degrees of freedom	*Significance level*		
	5%	*1%*	*0.1%*
1	0.996 92	0.999 877	0.999 9988
2	0.950 00	0.990 000	0.999 00
3	0.8793	0.958 73	0.991 16
4	0.8114	0.917 20	0.974 06
5	0.7545	0.8745	0.950 74
6	0.7076	0.8343	0.924 93
7	0.6664	0.7977	0.8982
8	0.6319	0.7646	0.8721
9	0.6021	0.7348	0.8471
10	0.5760	0.7079	0.8233
11	0.5529	0.6835	0.8010
12	0.5324	0.6614	0.7800
13	0.5139	0.6411	0.7603
14	0.4973	0.6226	0.7420
15	0.4821	0.6055	0.6524
16	0.4683	0.5897	0.7084
17	0.4555	0.5751	0.6932
18	0.4438	0.5614	0.6787
19	0.4329	0.5487	0.6652
20	0.4427	0.5368	0.6524
25	0.3809	0.4869	0.6974
30	0.3494	0.4487	0.5541
35	0.3246	0.4182	0.5189
40	0.3044	0.3932	0.4896
45	0.2875	0.3721	0.4648
50	0.2732	0.3541	0.4433
60	0.2500	0.3248	0.4078
70	0.2319	0.3017	0.3799
80	0.2172	0.2830	0.3568
90	0.2050	0.2673	0.3375
100	0.1946	0.2540	0.3211

Table S3: Critical values for the χ^2 statistic

Critical values of χ^2 at the 5%, 1% and 0.1% significance levels. Reject the null hypothesis if your value of χ^2 is larger than the tabulated value at the chosen significance level, for the calculated number of degrees of freedom.

Degrees of freedom	*Significance level*		
	5%	*1%*	*0.1%*
1	3.841	6.653	10.827
2	5.991	9.210	13.815
3	7.815	11.345	16.266
4	9.488	13.277	18.467
5	11.070	15.086	20.515
6	12.592	16.812	22.457
7	14.067	18.457	24.322
8	15.507	20.090	26.125
9	16.919	21.666	27.877
10	18.307	23.209	29.588
11	19.675	24.725	31.264
12	21.026	26.217	32.909
13	22.362	27.688	34.528
14	23.685	29.141	36.123
15	24.996	30.578	37.697
16	26.296	32.000	39.252
17	27.587	33.409	40.792
18	28.869	34.805	42.312
19	30.144	36.191	43.820
20	31.410	37.566	45.315
21	32.671	38.932	46.797
22	33.924	40.289	48.268
23	35.172	41.638	49.728
24	36.415	42.980	51.179
25	37.652	44.314	52.260
26	38.885	45.642	54.052
27	40.113	46.963	55.476
28	41.337	48.278	56.893
29	42.557	49.588	58.302
30	43.773	50.892	59.703

Index

LLYFRGELL COLEG MENAI LIBRARY
SAFLE FFRIDDOEDD SITE
BANGOR GWYNEDD LL57 2TP